FLOOD RISK MANAGEMENT: RESEARCH AND PRACTICE

PROCEEDINGS OF THE EUROPEAN CONFERENCE ON FLOOD RISK MANAGEMENT RESEARCH INTO PRACTICE (FLOOD*RISK* 2008), OXFORD, UK, 30 SEPTEMBER–2 OCTOBER 2008

Flood Risk Management: Research and Practice

Editors

Paul Samuels
Water Management, HR Wallingford, Wallingford, Oxfordshire, UK

Stephen Huntington
HR Wallingford Group, Wallingford, Oxfordshire, UK

William Allsop
Coastal Structures, HR Wallingford, Wallingford, Oxfordshire, UK

Jackie Harrop
HR Wallingford, Wallingford, Oxfordshire, UK

CRC Press
Taylor & Francis Group
Boca Raton   London   New York   Leiden

CRC Press is an imprint of the
Taylor & Francis Group, an **informa** business

A BALKEMA BOOK

CRC Press/Balkema is an imprint of the Taylor & Francis Group, an informa business

© 2009 Taylor & Francis Group, London, UK
Improving the understanding of the risk from groundwater flooding in the UK by D.M.J. Macdonald,
J.P. Bloomfield, A.G. Hughes, A.M. MacDonald, B. Adams & A.A. McKenzie
© British Geological Survey
The worst North Sea storm surge for 50 years: Performance of the forecasting system and implications for decision makers by K.J. Horsburgh, J. Williams, J. Flowerdew, K. Mylne, S. Wortley
© Crown copyright

Typeset by Vikatan Publishing Solutions (P) Ltd., Chennai, India.
Printed and bound in Great Britain by Antony Rowe (A CPI-group Company), Chippenham, Wiltshire.

All rights reserved. No part of this publication or the information contained herein may be reproduced, stored in a retrieval system, or transmitted in any form or by any means, electronic, mechanical, by photocopying, recording or otherwise, without written prior permission from the publisher.

Although all care is taken to ensure integrity and the quality of this publication and the information herein, no responsibility is assumed by the publishers nor the author for any damage to the property or persons as a result of operation or use of this publication and/or the information contained herein.

Published by: CRC Press/Balkema
    P.O. Box 447, 2300 AK Leiden, The Netherlands
    e-mail: Pub.NL@taylorandfrancis.com
    www.crcpress.com – www.taylorandfrancis.co.uk – www.balkema.nl

ISBN: 978-0-415-48507-4 (Hbk + CD-rom)

Table of contents

Foreword — XIX

Committees — XXI

KEYNOTE PRESENTATION

Coastal flooding: A view from a practical Dutchman on present and future strategies — 3
J.W. van der Meer

TECHNICAL PRESENTATIONS

Inundation modelling

Recent development and application of a rapid flood spreading method — 15
J. Lhomme, P. Sayers, B. Gouldby, P. Samuels, M. Wills & J. Mulet-Marti

Hydrodynamic modelling and risk analysis in RAMFLOOD project — 16
E. Bladé, M. Gómez-Valentín, J. Dolz, M. Sánchez-Juny, J. Piazzese, E. Oñate & G. Corestein

Testing and application of a practical new 2D hydrodynamic model — 17
J. Gutierrez Andres, J. Lhomme, A. Weisgerber, A. Cooper, B. Gouldby & J. Mulet-Marti

Floods study through coupled numerical modeling of 2D surface and sewage network flows — 18
C. Coulet, L. Evaux & A. Rebaï

Modelling of flooding and analysis of pluvial flood risk – demo case of UK catchment — 19
J.P. Leitão, S. Boonya-aroonnet, Č. Maksimović, R. Allitt & D. Prodanović

An integrated approach to modelling surface water flood risk in urban areas — 21
J.B. Butler, D.M. Martin, E.M. Stephens & L. Smith

Estimation of flood inundation probabilities using global hazard indexes based on hydrodynamic variables — 22
G.T. Aronica, P. Fabio, A. Candela & M. Santoro

Flood modeling for risk evaluation – a MIKE FLOOD vs. SOBEK 1D2D benchmark study — 23
P. Vanderkimpen, E. Melger & P. Peeters

Comparing forecast skill of inundation models of differing complexity: The case of Upton upon Severn — 25
K. Srinivas, M. Werner & N. Wright

Comparison of varying complexity numerical models for the prediction of flood inundation in Greenwich, UK — 26
T.J. Fewtrell, P.D. Bates, A. de Wit, N. Asselman & P. Sayers

Fast 2D floodplain modeling using computer game technology — 27
R. Lamb, A. Crossley & S. Waller

Grid resolution dependency in inundation modelling: A case study   29
S. Néelz & G. Pender

2D overland flow modelling using fine scale DEM with manageable runtimes   30
J.N. Hartnack, H.G. Enggrob & M. Rungø

Detailed 2D flow simulations as an onset for evaluating socio-economic impacts of floods   31
B.J. Dewals, S. Detrembleur, P. Archambeau, S. Erpicum & M. Pirotton

Ensemble Prediction of Inundation Risk and Uncertainty arising from Scour
(EPIRUS): An overview   33
*Q. Zou, D. Reeve, I. Cluckie, S. Pan, M.A. Rico-Ramirez, D. Han, X. Lv,
A. Pedrozo-Acuña & Y. Chen*

Flood risk assessment using broad scale two-dimensional hydraulic modelling – a case study
from Penrith, Australia   35
H. Rehman, R. Thomson & R. Thilliyar

Modelling and analysis of river flood impacts on sewage networks in urban areas   36
A. Kron, P. Oberle, A. Wetzel & N. Ettrich

Coastal flood risk modelling in a data rich world   38
R.D. Williams, M.R. Lawless & J. Walker

A multi-scale modelling procedure to quantify effects of upland land management on flood risk   40
*H.S. Wheater, B.M. Jackson, O. Francis, N. McIntyre, M. Marshall, I. Solloway,
Z. Frogbrook & B. Reynolds*

Updating flood maps using 2D models in Italy: A case study   41
F. Nardi, J.S. O'Brien, G. Cuomo, R. Garcia & S. Grimaldi

Real-time validation of a digital flood-inundation model: A case-study from Lakes Entrance,
Victoria, Australia   43
P.J. Wheeler, J. Kunapo, M.L.F. Coller, J.A. Peterson & M. McMahon

Dispelling the myths of urban flood inundation modelling   44
D. Fortune

Flood risk in urban areas caused by levee breaching   45
A. Paquier, C. Peyre, N. Taillefer & M. Chenaf

RISK-EOS flood risk analysis service for Europe   46
V. Holzhauer, M. Müller & A. Assmann

Flood inundation modelling: Model choice and application   47
*N. Asselman, J. ter Maat, A. de Wit, G. Verhoeven, S. Soares Frazão, M. Velickovic, L. Goutiere,
Y. Zech, T. Fewtrell & P. Bates*

Risk maps of torrential rainstorms   48
A. Assmann, M. Krischke & E. Höppner

Decision Support System for flood forecasting and risk mitigation in the context of
Romanian water sector   49
I. Popescu, A. Jonoski & A. Lobbrecht

Developing a rapid mapping and monitoring service for flood management using remote
sensing techniques   50
V. Craciunescu, C. Flueraru, G. Stancalie & A. Irimescu

A framework for Decision Support Systems for flood event management – application to the
Thames and the Schelde Estuaries   51
D.M. Lumbroso, M.J.P. Mens & M.P. van der Vat

Modelling tsunami overtopping of a sea defence by shallow-water Boussinesq, VOF
and SPH methods 52
P. Stansby, R. Xu, B. Rogers, A. Hunt, A. Borthwick & P. Taylor

Modelling the 2005 Carlisle flood event using LISFLOOD-FP and TRENT 54
*J.C. Neal, P.D. Bates, T.J. Fewtrell, N.G. Wright, I. Villanueva, N.M. Hunter &
M.S. Horritt*

Experience of 1D and 2D flood modelling in Australia – a guide to model selection based
on channel and floodplain characteristics 55
J.M. Hannan & J. Kandasamy

Computationally efficient flood water level prediction (with uncertainty) 56
K. Beven, P. Young, D. Leedal & R. Romanowicz

Optimization of 2D flood models by semi-automated incorporation of flood diverting
landscape elements 57
P. Vanderkimpen, P. Peeters & K. Van der Biest

Understanding the runoff response of the Ourthe catchment using spatial and temporal
characteristics of the storm field obtained by radar 59
P. Hazenberg, H. Leijnse, R. Uijlenhoet & L. Delobbe

The importance of spill conceptualizations and head loss coefficients in a quasi two-dimensional
approach for river inundation modelling 60
M.F. Villazón & P. Willems

Inundation scenario development for damage evaluation in polder areas 61
L.M. Bouwer, P. Bubeck, A.J. Wagtendonk & J.C.J.H. Aerts

System analysis

Importance of river system behaviour in assessing flood risk 65
M.C.L.M. van Mierlo, T. Schweckendiek & W.M.G. Courage

Development and evaluation of an integrated hydrological modelling tool for the Water Framework
Directive and Floods Directive 66
M.B. Butts, E. Fontenot, M. Cavalli, C.Y. Pin, T.S. Jensen, T. Clausen & A. Taylor

A comparison of modelling methods for urban flood risk assessment 67
C.J. Digman, T. Bamford, D.J. Balmforth, N.M. Hunter & S.G. Waller

Coastal flood risk analysis driven by climatic and coastal morphological modelling 68
M.J. Walkden, J.W. Hall, R. Dawson, N. Roche & M. Dickson

Micro-scale analysis of flood risk at the German Bight Coast 69
G. Kaiser, S.D. Hofmann, H. Sterr & A. Kortenhaus

Flood hazard mapping for coastal storms in the Delta Ebro 70
D. Alvarado-Aguilar & J.A. Jiménez

RAMWASS Decision Support System (DSS) for the risk assessment
of water-sediment-soil systems – application of a DSS prototype to a test site
in the lower part of the Elbe river valley, Germany 71
B. Koppe, B. Llacay & G. Peffer

Radar based nowcasting of rainfall events – analysis and assessment
of a one-year continuum 72
H.-R. Verworn & S. Krämer

On the quality of Pareto calibration solutions of conceptual rainfall-runoff models   73
A.-R. Nazemi, A.H. Chan, A. Pryke & X. Yao

Model reuse and management in flood risk modelling   74
R. Khatibi

International programmes

Flood Risk from Extreme Events (FREE): A NERC-directed research programme – understanding the science of flooding   77
C.G. Collier

Advances in flood risk management from the FLOODsite project   78
P.G. Samuels, M.W. Morris, P. Sayers, J-D. Creutin, A. Kortenhaus, F. Klijn, E. Mosselman, A. van Os & J. Schanze

The Tyndall Centre Coastal Simulator and Interface (CoastS)   79
R.J. Nicholls, M. Mokrech, S.E. Hanson, P. Stansby, N. Chini, M. Walkden, R. Dawson, N. Roche, J.W. Hall, S.A. Nicholson-Cole, A.R. Watkinson, S.R. Jude, J.A. Lowe, J. Leake, J. Wolf, C. Fontaine, M. Rounsvell & L. Acosta-Michlik

The social impacts of flooding in Scotland: A national and local analysis   81
A. Werritty, D.M. Houston, M. Jobe, T. Ball, A.C.W. Tavendale & A.R. Black

The Flood Risk Management Research Consortium (FRMRC)   82
I.D. Cluckie

EIB financing for flood risk mitigation   83
C. Gleitsmann

One nation, one policy, one program flood risk management   84
P.D. Rabbon, L.J. Zepp & J.R. Olsen

Toward a transnational perspective on flood-related research in Europe – experiences from the CRUE ERA-Net   86
A. Pichler, V. Jackson, S. Catovsky & T. Deppe

Infrastructure and assets

Hazards from wave overtopping   89
W. Allsop, T. Bruce, T. Pullen & J. van der Meer

Time-dependent reliability analysis of anchored sheet pile walls   91
F.A. Buijs, P.B. Sayers, J.W. Hall & P.H.A.J.M. van Gelder

Analysis of tsunami hazards by modelling tsunami wave effects   93
T. Rossetto, W. Allsop, D. Robinson, I. Chavet & P.-H. Bazin

Influence of management and maintenance on erosive impact of wave overtopping on grass covered slopes of dikes; Tests   95
G.J. Steendam, W. de Vries, J.W. van der Meer, A. van Hoven, G. de Raat & J.Y. Frissel

Sea wall or sea front? Looking at engineering for Flood and Coastal Erosion Risk Management through different eyes   97
J. Simm

The new turner contemporary gallery – an example of an urban coastal flood risk assessment   98
H. Udale-Clarke, W. Allsop, P. Hawkes & P. Round

EurOtop – overtopping and methods for assessing discharge 99
T. Pullen, N.W.H. Allsop, T. Bruce, A. Kortenhaus, H. Schüttrumpf & J.W. van der Meer

Reliable prediction of wave overtopping volumes using Bayesian neural networks 101
G.B. Kingston, D.I. Robinson, B.P. Gouldby & T. Pullen

Calculation of fragility curves for flood defence assets 102
J.W. van der Meer, W.L.A. ter Horst & E.H. van Velzen

Reservoir flood risk in the UK 104
A.L. Warren

Modelling breach initiation and growth 105
M.W. Morris, M.A.A.M. Hassan, A. Kortenhaus, P. Geisenhainer, P.J. Visser & Y. Zhu

A probabilistic failure model for large embankment dams 107
N.P. Huber, J. Köngeter & H. Schüttrumpf

Reliability analysis of flood defence structures and systems in Europe 109
P. van Gelder, F. Buijs, W. ter Horst, W. Kanning, C. Mai Van, M. Rajabalinejad, E. de Boer, S. Gupta, R. Shams, N. van Erp, B. Gouldby, G. Kingston, P. Sayers, M. Wills, A. Kortenhaus & H.-J. Lambrecht

PCRIVER—software for probability based flood protection 111
U. Merkel, B. Westrich & A. Moellmann

Representing fragility of flood and coastal defences: Getting into the detail 112
J. Simm, B. Gouldby, P. Sayers, J-J. Flikweert, S. Wersching & M. Bramley

Application of 3D serious games in levee inspection education 113
M. Hounjet, J. Maccabiani, R. van den Bergh & C. Harteveld

Strategic appraisal of flood risk management options over extended timescales: Combining
scenario analysis with optimization 114
J.W. Hall, T.R. Phillips, R.J. Dawson, S.L. Barr, A.C. Ford, M. Batty, A. Dagoumas & P.B. Sayers

Embedding new science into practice – lessons from the development and application
of a Performance-based asset management system 116
C. Mitchell, O. Tarrant, D. Denness, P. Sayers, J. Simm & M. Bramley

Study of flood embankment behaviour induced by air entrapment 117
D. Lesniewska, H. Zaradny, P. Bogacz & J. Kaczmarek

Assessment of flood retention in polders using an interlinked one-two-dimensional hydraulic model 119
M. Kufeld, H. Schüttrumpf & D. Bachmann

Fragility curve calculation for technical flood protection measures by the Monte Carlo analysis 120
D. Bachmann, N.P. Huber & H. Schüttrumpf

Application of GMS system in the Czech Republic – practical use of IMPACT, FLOODSite
and GEMSTONE projects outcomes 121
Z. Boukalová & V. Beneš

Failure modes and mechanisms for flood defence structures 122
M.W. Morris, W. Allsop, F.A. Buijs, A. Kortenhaus, N. Doorn & D. Lesniewska

Non-structural approaches (CRUE project)

Flood risk map perception through experimental graphic semiology 127
S. Fuchs, W. Dorner, K. Spachinger, J. Rochman & K. Serrhini

Quantifying the benefits of non-structural flood risk management measures 128
R.J. Dawson, N. Roche, A.C. Ford, S.L. Barr, J.W. Hall, J. Werritty, T. Ball, A. Werritty, M. Raschke & K. Thürmer

Efficiency of non-structural flood mitigation measures: "room for the river" and "retaining water in the landscape" 130
S. Salazar, F. Francés, J. Komma, G. Blöschl, T. Blume, T. Francke & A. Bronstert

Flood risk reduction by PReserving and restOring river FLOODPLAINs – PRO_FLOODPLAIN 131
H. Habersack, C. Hauer, B. Schober, E. Dister, I. Quick, O. Harms, M. Wintz, E. Piquette & U. Schwarz

The use of non structural measures for reducing the flood risk in small urban catchments 132
E. Pasche, N. Manojlovic, D. Schertzer, J.F. Deroubaix, I. Tchguirinskaia, E. El Tabach, R. Ashley, R. Newman, I. Douglas, N. Lawson & S. Garvin

EWASE—Early Warning Systems Efficiency: Evaluation of flood forecast reliability 134
K. Schröter, M. Ostrowski, M. Gocht, B. Kahl, H.-P. Nachtnebel, C. Corral & D. Sempere-Torres

Flood risk assessment in an Austrian municipality comprising the evaluation of effectiveness and efficiency of flood mitigation measures 135
C. Neuhold & H.-P. Nachtnebel

EWASE—Early Warning Systems Efficiency – risk assessment and efficiency analysis 136
M. Gocht, K. Schröter, M. Ostrowski, C. Rubin & H.P. Nachtnebel

Flood risk management strategies in European Member States considering structural and non-structural measures 138
J. Schanze, G. Hutter, E. Penning-Rowsell, D. Parker, H.-P. Nachtnebel, C. Neuhold, V. Meyer & P. Königer

Long term planning, integrated portfolios & spatial planning

The OpenMI-LIFE project – putting integrated modelling into practice in flood management 141
D. Fortune

A method for developing long-term strategies for flood risk management 142
K.M. de Bruijn, M.J.P. Mens & F. Klijn

Flood Risk Mapping, using spatially based Systems Engineering 143
R. Raaijmakers

Finding a long term solution to flooding in Oxford: The challenges faced 144
L.G.A. Ball, M.J. Clegg, L. Lewis & G. Bell

Risk analysis and decision-making for optimal flood protection level in urban river management 145
M. Morita

An integrated risk-based multi criteria decision-support system for flood protection measures in riversheds—REISE 146
N.P. Huber, D. Bachmann, H. Schüttrumpf, J. Köngeter, U. Petry, M. Pahlow, A.H. Schumann, J. Bless, G. Lennartz, O. Arránz-Becker, M. Romich & J. Fries

Integrated methodologies for flood risk management practice in European pilot sites 148
J. Schanze, P. Bakonyi, M. Borga, B. Gouldby, M. Marchand, J.A. Jiménez & H. Sterr

Underpinning flood risk management: A digital terrain model for the 21st century 150
M. Stileman & D. Henderson

Integrated land and water management in floodplains in England 151
H. Posthumus, J.R. Rouquette, J. Morris, T.M. Hess, D.J. Gowing & Q.L. Dawson

Putting people and places at the centre: Improving institutional and social responses to flooding 153
C. Twigger-Ross, A. Fernandez-Bilbao, L. Colbourne, S. Tapsell, N. Watson, E. Kashefi,
G. Walker & W. Medd

Delivering Integrated Urban Drainage – current obstacles and a proposed
SUDS planning support tool 154
V.R. Stovin, S.L. Moore, S.H. Doncaster & B. Morrow

Strategic planning for long-term Flood Risk Management – findings from case studies
in Dresden and London 155
G. Hutter & L. McFadden

Extreme flood events & flood management strategy at the Slovak-Austrian part of the
Morava river basin 156
M. Lukac & K. Holubova

Using non-structural responses to better manage flood risk in Glasgow 157
R. Ashley, R. Newman, F. McTaggart, S. Gillon, A. Cashman, G. Martin & S. Molyneux-Hodgson

Vulnerability and resilience, human and social impacts

The policy preferences of citizens, scientists and policy makers 161
J.H. Slinger, M. Cuppen & M. Marchand

Analysis of the human and social impacts of flooding in Carlisle 2005 and Hull 2007 162
P. Hendy

Institutional and social responses to flooding from a resilience perspective 163
N. Watson, E. Kashefi, W. Medd, G. Walker, S. Tapsell & C. Twigger-Ross

Flood, vulnerability and resilience: A real-time study of local recovery following the floods
of June 2007 in Hull 164
R. Sims, W. Medd, E. Kashefi, M. Mort, N. Watson, G. Walker & C. Twigger-Ross

Increasing resilience to storm surge flooding: Risks, social networks and local champions 165
H. Deeming

A new model to estimate risk to life for European flood events 166
S.M. Tapsell, S.J. Priest, T. Wilson, C. Viavattene & E.C. Penning-Rowsell

Towards flood risk management with the people at risk: From scientific analysis to practice
recommendations (and back) 167
A. Steinführer, C. Kuhlicke, B. De Marchi, A. Scolobig, S. Tapsell & S. Tunstall

Use of human dimensions factors in the United States and European Union 168
S. Durden & C.M. Dunning

Double whammy? Are the most at risk the least aware? A study of environmental justice
and awareness of flood risk in England and Wales 169
J.L. Fielding

Improving public safety in the United States – from Federal protection to shared
flood risk reduction 170
E.J. Hecker, L.J. Zepp & J.R. Olsen

Evaluating the benefits and limitations of property based flood resistance
and resilience – a UK perspective 171
N. Thurston, B. Finlinson, N. Williams, J. Shaw, J. Goudie & T. Harries

Flood risk management: Experiences from the Scheldt Estuary case study 172
M. Marchand, K.M. de Bruijn, M.J.P. Mens, J.H. Slinger, M.E. Cuppen,
J. Krywkow & A. van der Veen

Overcoming the barriers to household-level adaptation to flood risk 173
T. Harries

Human vulnerability to flash floods: Addressing physical exposure and behavioural questions 174
I. Ruin, J.-D. Creutin, S. Anquetin, E. Gruntfest & C. Lutoff

Assessment of extremes

Estimating extremes in a flood risk context. The FLOODsite approach 177
A. Sanchez-Arcilla, D. Gonzalez-Marco & P. Prinos

Inter-site dependence in extremes: Unlocking extra information 178
D.W. Reed

The Flood Estimation Handbook and UK practice: Past, present and future 179
E.J. Stewart, T.R. Kjeldsen, D.A. Jones & D.G. Morris

Extreme precipitation mapping for flood risk assessment in ungauged basins
of the upper Hron River basin in Slovakia 180
S. Kohnová, J. Szolgay, K. Hlavčová, L. Gaál & J. Parajka

River flood frequency approaches for ungauged sites 181
A. Calver & E.J. Stewart

Non-stationary point process models for extreme storm surges 182
P. Galiatsatou & P. Prinos

Bayesian non-parametric quantile regression using splines for modelling wave heights 183
P. Thompson, D. Reeve, J. Stander, Y. Cai & R. Moyeed

Multiscale probabilistic risk assessment 185
C. Keef, R. Lamb, P. Dunning & J.A. Tawn

Improving the understanding of the risk from groundwater flooding in the UK 186
*D.M.J. Macdonald, J.P. Bloomfield, A.G. Hughes, A.M. MacDonald, B. Adams &
A.A. McKenzie*

Radar observation of storm rainfall for flash-flood forecasting 187
*G. Delrieu, A. Berne, M. Borga, B. Boudevillain, B. Chapon,
P.-E. Kirstetter, J. Nicol, D. Norbiato & R. Uijlenhoet*

Climate change impact on hydrological extremes along rivers in Belgium 189
O.F. Boukhris & P. Willems

Uncertainties in 1D flood level modeling: Stochastic analysis of upstream discharge
and friction parameter influence 190
N. Goutal, P. Bernardara, E. de Rocquigny & A. Arnaud

Civil contingency, emergency planning, flood event management

Reservoir safety in England and Wales – reducing risk, safeguarding people 193
I.M. Hope & A.K. Hughes

A comparison of evacuation models for flood event management – application
on the Schelde and Thames Estuaries 194
M.J.P. Mens, M. van der Vat & D. Lumbroso

Hydrodynamic and loss of life modelling for the 1953 Canvey Island flood 195
M. Di Mauro & D. Lumbroso

Short-range plain flood forecasting and risk management in the Bavarian Danube basin 196
M. Mueller, M. Tinz, A. Assmann, P. Krahe, C. Rachimow, K. Daamen, J. Bliefernicht,
C. Ebert, M. Kunz, J.W. Schipper, G. Meinel & J. Hennersdorf

Fast access to ASAR imagery for rapid mapping of flood events 198
R. Cossu, Ph. Bally, O. Colin, E. Schoepfer & G. Trianni

Benefits of 2D modelling approach for urban flood management 199
E. David, M. Erlich & A. Masson

Computer modelling of hydrodynamic conditions on the Lower Kuban under various scenarios
and definition of limiting values of releases from the Krasnodar, Shapsugsky and Varnavinsky
hydrounits for prevention of flooding 200
M.A. Volinov, A.L. Buber, M.V. Troshina, A.M. Zeiliguer & O.S. Ermolaeva

Flood warning in the UK: Shifting the focus 202
C.L. Twigger-Ross, A. Fernandez-Bilbao, G.P. Walker, H. Deeming, E. Kasheri,
N. Watson & S. Tapsell

New approaches to ex-post evaluation of risk reduction measures: The example of flood
proofing in Dresden, Germany 203
A. Olfert & J. Schanze

Dilemmas in land use planning in flood prone areas 204
A. Scolobig & B. De Marchi

Emergency management of flood events in Alpine catchments 205
H. Romang & C. Wilhelm

Flood forecasting and warning

Flood warning in smaller catchments 209
H. Romang, F. Dufour, M. Gerber, J. Rhyner, M. Zappa, N. Hilker & C. Hegg

A prototype of road warning system in flood prone area 210
P.-A. Versini, E. Gaume & H. Andrieu

Snow and glacier melt – a distributed energy balance model within a flood forecasting system 211
J. Asztalos, R. Kirnbauer, H. Escher-Vetter & L. Braun

Analysis of weather radar and rain gauges for flood forecasting 212
M.T.J. Bray, D. Han, I. Cluckie & M. Rico-Ramirez

Integration of hydrological information and knowledge management for rapid decision-making
within European flood warning centres 213
F. Schlaeger, D. Witham & R. Funke

Local warning systems in Slovakia 215
D. Lešková, D. Kyselová, P. Rončák & M. Hollá

The provision of site specific flood warnings using wireless sensor networks 216
P. Smith, K. Beven, W. Tych, D. Hughes, G. Coulson & G. Blair

Managing flood risk in Bristol, UK – a fluvial & tidal combined forecasting challenge 217
M. Dale, O. Pollard, K. Tatem & A. Barnes

Off-line flood warning concept for railways 218
U. Drabek, T. Nester & R. Kirnbauer

Satellite observation of storm rainfall for flash-flood forecasting in small and medium-size basins 219
C. Görner, N. Jatho, C. Bernhofer & M. Borga

Potential warning services for groundwater and pluvial flooding 220
D. Cobby, R. Falconer, G. Forbes, P. Smyth, N. Widgery, G. Astle, J. Dent & B. Golding

Data assimilation and adaptive real-time forecasting of water levels
in the river Eden catchment, UK 221
D. Leedal, K. Beven, P. Young & R. Romanowicz

To which extent do rainfall estimation uncertainties limit the accuracy of flash flood forecasts? 222
L. Moulin, E. Gaume & Ch. Obled

Advances in radar-based flood warning systems. The EHIMI system and the experience
in the Besòs flash-flood pilot basin 223
C. Corral, D. Velasco, D. Forcadell, D. Sempere-Torres & E. Velasco

Flash flood risk management: Advances in hydrological forecasting and warning 224
M. Borga, J.-D. Creutin, E. Gaume, M. Martina, E. Todini & J. Thielen

Decision support system for flood forecasting in the Guadalquivir river basin 225
L. Rein, A. Linares, E. García & A. Andrés

Operational flash flood forecasting chain using hydrological
and pluviometric precursors 226
G. Brigandì & G.T. Aronica

Online updating procedures for flood forecasting with a continuous rainfall-runoff-model 227
B. Kahl & H.P. Nachtnebel

GIS technology in water resources parameter extraction in flood forecasting 228
V. Ramani Bai, G. Ramadas & R. Simons

Combining weather radar and raingauge data for hydrologic applications 230
C. Mazzetti & E. Todini

The worst North Sea storm surge for 50 years: Performance of the forecasting system
and implications for decision makers 231
K.J. Horsburgh, J. Williams, J. Flowerdew, K. Mylne & S. Wortley

Probabilistic coastal flood forecasting 232
P.J. Hawkes, N.P. Tozer, A. Scott, J. Flowerdew, K. Mylne & K. Horsburgh

Coastal flood inundation modelling for North Sea lowlands 234
S. Burg, F. Thorenz & H. Blum

New north east of England tidal flood forecasting system 235
A. Lane, K. Hu, T.S. Hedges & M.T. Reis

Impact of extreme waves and water levels in the south Baltic Sea 236
H. Hanson & M. Larson

Bayesian rainfall thresholds for flash flood guidance 238
M.L.V. Martina & E. Todini

Environmental impacts, morphology & sediments

Assessment of hydraulic, economic and ecological impacts of flood polder
management – a case study from the Elbe River, Germany 241
S. Förster & A. Bronstert

Development of estuary morphology models 242
J.M. Huthnance, A. Lane, H. Karunarathna, A.J. Manning, D.E. Reeve, P.A. Norton, A.P. Wright,
R.L. Soulsby, J. Spearman, I.H. Townend & S. Surendran

A GIS-based risk assessment methodology for flood pollutants 244
A. Sauer, J. Schanze & U. Walz

Environmental impact of flash floods in Hungary 245
S. Czigány, E. Pirkhoffer & I. Geresdi

Predicting beach morphology as part of flood risk assessment 246
J.M. Horrillo-Caraballo & D.E. Reeve

Alkborough scheme reduces extreme water levels in the Humber Estuary and
creates new habitat 248
D. Wheeler, S. Tan, N. Pontee & J. Pygott

Managing coastal change: Walberswick to Dunwich 249
M. Cali, A. Parsons, N. Pontee, L. Batty, S. Duggan & P. Miller

Uncertainties in the parameterisation of rainfall-runoff-models to quantify land-use effects
in flood risk assessment 250
A. Wahren, K.H. Feger, H. Frenzel & K. Schwärzel

Impact of the barrage construction on the hydrodynamic process in the severn estuary using
a 2D finite volume model 253
J. Xia, R.A. Falconer & B. Lin

Risk sharing, equity and social justice

From knowledge management to prevention strategies: The example of the tools developed
by French insurers 257
J. Chemitte & R. Nussbaum

What's 'fair' about flood and coastal erosion risk management? A case study evaluation
of policies and attitudes in England 258
C. Johnson, S. Tunstall, S. Priest, S. McCarthy & E. Penning-Rowsell

Flood risk perceptions in the Dutch province of Zeeland: Does the public still support
current policies? 259
J. Krywkow, T. Filatova & A. van der Veen

A partnership approach – public flood risk management and private insurance 260
M. Crossman, S. Surminski, A. Philp & D. Skerten

The international teaching module FLOODmaster – an integrated part of a European educational
platform on flood risk management 262
J. Seegert, C. Bernhofer, K. Siemens & J. Schanze

Decision support for strategic flood risk planning – a generic conceptual model 263
A.G.J. Dale & M.V.T. Roberts

Who benefits from flood management policies? 264
N. Walmsley, E. Penning-Rowsell, J. Chatterton & K. Hardy

Uncertainty

Long term planning – robust strategic decision making in the face of gross uncertainty
(tools and application to the Thames) 267
C. Mc Gahey & P.B. Sayers

Anticipatory water management for advanced flood control 268
S.J. van Andel, A.H. Lobbrecht & R.K. Price

Staged uncertainty and sensitivity analysis within flood risk analysis 269
B. Gouldby & G. Kingston

Assessing uncertainty in rainfall-runoff models: Application of data-driven models 270
D.L. Shrestha & D.P. Solomatine

Flash floods

European flash floods data collation and analysis 273
V. Bain, O. Newinger, E. Gaume, P. Bernardara, M. Barbuc, A. Bateman, J. Garcia, V. Medina, D. Sempere-Torres, D. Velasco, L. Blaškovičová, G. Blöschl, A. Viglione, M. Borga, A. Dumitrescu, A. Irimescu, G. Stancalie, S. Kohnova, J. Szolgay, A. Koutroulis, I. Tsanis, L. Marchi & E. Preciso

Representative flash flood events in Romania Case studies 275
G. Stancalie, B. Antonescu, C. Oprea, A. Irimescu, S. Catana, A. Dumitrescu, M. Barbuc & S. Matreata

Changes in flooding pattern after dam construction in Zadorra river (Spain): The events of October 1953 and February 2003 276
A. Ibisate

Post flash flood field investigations and analyses: Proposal of a methodology and illustrations of its application 277
E. Gaume & M. Borga

Hydrological and hydraulic analysis of the flash flood event on 25 October 2007 in North-Eastern part of Sicily, Italy 278
G.T. Aronica, G. Brigandì, C. Marletta & B. Manfrè

The day roads became rivers: A GIS-based assessment of flash floods in Worcester 279
F. Visser

Risk and economic assessments

Flood risk mapping of Austrian railway lines 283
A. Schöbel, A.H. Thieken & R. Merz

Correlation in time and space: Economic assessment of flood risk with the Risk Management Solutions (RMS) UK River Flood Model 284
D. Lohmann, S. Eppert, A. Hilberts, C. Honegger & A. Steward-Menteth

A case study of the Thames Gateway: Flood risk, planning policy and insurance loss potential 285
J. Eldridge & D.P. Horn

Integration of accurate 2D inundation modelling, vector land use database and economic damage evaluation 286
J. Ernst, B.J. Dewals, P. Archambeau, S. Detrembleur, S. Erpicum & M. Pirotton

Planning for flood damages reduction: A case study 287
M. Karamouz, A. Moridi & A. Ahmadi

High resolution inundation modelling as part of a multi-hazard loss modelling tool 288
S. Reese & G. Smart

Estimation of flood losses due to business interruption 289
I. Seifert, H. Kreibich, B. Merz & A. Thieken

Residential flood losses in Perth, Western Australia 290
M.H. Middelmann

A multicriteria flood risk assessment and mapping approach *V. Meyer, D. Haase & S. Scheuer*	291
New developments in maximizing flood warning response and benefit strategies *S.J. Priest, D.J. Parker & S. Tapsell*	292
Development of a damage and casualties tool for river floods in northern Thailand *J.K. Leenders, J. Wagemaker, A. Roelevink, T.H.M. Rientjes & G. Parodi*	293
Synthetic water level building damage relationships for GIS-supported flood vulnerability modeling of residential properties *M. Neubert, T. Naumann & C. Deilmann*	294
Impacts of the summer 2007 floods on agriculture in England *H. Posthumus, J. Morris, T.M. Hess, P. Trawick, D. Neville, E. Phillips & M. Wysoki*	295

Climate change

Simulating flood-peak probability in the Rhine basin and the effect of climate change *A.H. te Linde & J.C.J.H. Aerts*	299
Climate changes in extreme precipitation events in the Elbe catchment of Saxony *C. Görner, J. Franke, C. Bernhofer & O. Hellmuth*	300
A methodology for adapting local drainage to climate change *R.M. Ashley, J.R. Blanksby, A. Cashman & R. Newman*	301
Exploring and evaluating futures of riverine flood risk systems – the example of the Elbe River *J. Luther & J. Schanze*	303
Author index	305

Foreword

Since the dawn of civilisation human society has been shaped by its interaction with water—whether too much or too little. Indeed, water is one of the powerful forces of nature that have formed the planet on which we live. Floods are not new—they continually make news because of both tragic effects on individuals and acts of heroism in the emergency. The impact of floods on people can be dramatic even if there are no fatalities; pictures of rescue make good television material with stories told of bravery and fortunate escapes. However, the aftermath of a flood is distressing with personal possessions ruined, houses deep in sewage-contaminated mud, vital services disrupted and businesses destroyed. Cleaning up, community recovery and economic restoration takes months and for many there remains the fear that the next time it rains or a severe storm is predicted, the experience will be repeated.

When discussing flood risk it is important to remember that "risk" is entirely a human concern; floods from river, estuary or coast are predominantly natural events; they are random. The risk arises because the human use and value of the river and coastal plains conflicts with their natural functions of storage and movement of water during a flood. Of course, the potential causes for some flooding are man-made, for example following the breaching of a dam or a flood embankment, and some floods are triggered by other hazards such as tsunami following an earthquake. In many respects the types of impact on people are similar to the more common sources of flooding—but probably they are more severe as these are often less predictable events. Catalogues of recent disasters are common in the introduction to volumes such as this, but we do not dwell here on recent floods, or on the effect of climate change, as there is much in the text.

It is, however, essential to comment on a major development in flood management policy from the European Union which will affect flood risk management in all EU Member States. On the 26th November 2007 the European Directive on the assessment and management of flood risks was enacted. This will be transposed into national legislation in each Member State within 2 years and sets out a set of actions on preliminary flood risk assessment, flood risk mapping and the preparation of flood risk management plans to be completed by the end of 2015. The Directive covers all sources of flooding (not just rivers, but coastal floods, urban and groundwater floods); it requires planning at basin scale and has specific requirements for trans-national basins; and, in all cases the potential impacts of climate change on the flood conditions need to be considered. The use of the phrase "management of flood risks" in the title of the Directive indicates that European policy has progressed away from a philosophy of flood control to the acceptance that flood risks should be managed.

FLOODrisk 2008 marks the completion of some substantial research projects:

- FLOODsite—an Integrated Project in the EC Sixth Framework Programme
- The first phase of the Flood Risk Management Research Consortium
- The first common call of the CRUE ERA-NET

FLOODsite is the largest ever EC research project on floods, and will be completed in early 2009. The FLOODsite consortium involves 37 of Europe's leading institutes and universities and brings together scientists from many disciplines along with public and private sector involvement from 13 countries. There are over 30 project tasks including the pilot applications in Belgium, the Czech Republic, France, Germany, Hungary, Italy, the Netherlands, Spain, and the UK. FLOODsite covers the physical, environmental, ecological and socio-economic aspects of floods from rivers, estuaries and the sea. In this volume there are papers on many aspects of the FLOODsite project.

The Flood Risk Management Research Consortium (FRMRC) was established in the UK by the Engineering and Physical Sciences Research Council to undertake an integrated programme of research to support effective flood risk management by

- establishing a programme of "cutting edge" research to enhance flood risk management practice worldwide;

- short-term delivery of tools and techniques to support short term improvements in flood risk management in the United Kingdom; and,
- development and training of the next generation of flood risk management professionals through their involvement in and exposure to the consortium's research.

The FRMRC involved over 20 UK universities with research in eight priority areas. FRMRC has recently completed the first phase of its multi-disciplinary programme of research; this volume contains several papers on their results. A second phase of the FRMRC programme commenced during 2008.

Whereas FRMRC and FLOODsite are both research projects, the CRUE ERA-NET does not directly carry out research. Rather it is a network of the major research funders in the EU who are exploring how to integrate their national research programmes more closely as part of the EU policy to strengthen the European Research Area. As part of this closer cooperation the CRUE partners have developed a vision for the future research needed and issued a common call for research on non-structural measures for flood risk management. The concept of the common call was to explore how national programmes with their different regulations could work together in identifying research topics and jointly tendering, commissioning, monitoring and evaluating research projects. The scientific advances from these first common call projects are presented within this volume.

In setting up FLOODrisk 2008 our intention was to cover flood risk management in an integrated and comprehensive way. Thus the call and selection of papers covered the physical and social sciences, included policy and practice, and ranged from long-term planning, emergency management and post-flood recovery. The theme of the conference is research into practice and we hope that much of the research discussed at FLOODrisk 2008 will improve the scientific evidence and practice in the actions of the Floods Directive in Europe and find application worldwide.

It is our pleasure to welcome you to the FLOODrisk 2008 conference.

Stephen Huntington
Chairman of the International Scientific Committee

Paul Samuels
Chairman of the Local Organising Committee

Committees

INTERNATIONAL SCIENTIFIC COMMITTEE

Professor Stephen Huntington (Chair)	HR Wallingford, UK
Dr Peter Bakonyi	VITUKI, Budapest, Hungary
Professor Eelco van Beek	Deltares, The Netherlands
Professor Marco Borga	Università di Padova, Italy
Professor Ian Cluckie	University of Bristol, UK
Professor Jean-Dominique Creutin	INP Grenoble, France
Dr Rolf Deigaard	DHI Group, Denmark
Ronnie Falconer	Jacobs / EWA, UK
Dr Frans Klijn	Deltares, The Netherlands
Dr Andreas Kortenhaus	Universität Braunschweig, Germany
Dr Elisabeth Lipiatou	DG Research, Brussels, EU
Dr Joan Pope	US Army Corps of Engineers, USA
Professor Panos Prinos	Aristotle University, Thessaloniki, Greece
Professor Paul Samuels	HR Wallingford, UK
Paul Sayers	HR Wallingford, UK
Professor Agustin Sanchez-Arcilla	Universitat Politècnica de Catalunya, Spain
Dr Patrick Sauvaget	SOGREAH, France
Dr Jochen Schanze	IÖR, Dresden, Germany
Professor Gheorge Stancalie	NMA, Bucharest, Romania

LOCAL ORGANISING COMMITTEE

Professor Paul Samuels	HR Wallingford, UK
Professor Garry Pender	Heriot-Watt University, UK
Professor William Allsop	HR Wallingford, UK
Jackie Harrop	HR Wallingford, UK
Chris Grandy	Creative Conferences, UK
Sue Frye	Creative Conferences, UK

KEYNOTE PRESENTATION

Coastal flooding: A view from a practical Dutchman on present and future strategies

J.W. van der Meer
Van der Meer Consulting BV, Heerenveen, The Netherlands

ABSTRACT: This key note paper intends to feed further discussion on safety against coastal flooding. It will mainly be based on the Dutch situation, where half of the population lives below sea level, but the paper will give enough discussion points for other situations. Observations, conclusions, etc., made in this paper and the presentation are on the personal account of the author, so do not represent any official view from the Netherlands.

The paper briefly describes the history of creating safety against flooding, which started after the large flooding in 1953 in the south west of the Netherlands with almost 2000 casualties. This led to the situation with high and strong dikes, which should withstand a storm with a certain return period between 2,500 and 10,000 years. A discussion started already 15 years ago on how to derive new rules, based on probability of flooding or even on flood risk. This discussion continues, but is now fed with many more calculations on failure of flood defence assets, breaching, inundation, damage, evacuation and last but not least: indestructible dikes.

1 INTRODUCTION

Coastal flooding has always been an important issue in the Netherlands, mainly because the whole country covers more or less the delta of the rivers Rhine and Meuse, and river delta's are by definition low compared to the sea. By protecting the low lying areas with dikes, the areas themselves settled by a metre or more and became even lower than the natural delta. Protection against flooding became more and more important.

The driving force for coastal flooding in the Netherlands will always be a very severe storm. In other countries also hurricanes or tsunamis may be driving forces. River flooding in the Netherlands, however, is closely linked to coastal flooding, mainly for two reasons. First there are estuarine areas where both a storm or a high river discharge may give flooding. The second reason is that the whole safety system in the Netherlands is not separated in coastal or river flooding, but is simply based on flooding in general. The paper will discuss some items where this may lead to wrong interpretations, basically due to not understanding fully the difference between the two driving forces, severe storm or high river discharge.

The word "dike" means any structure made out of soil (sand, clay), often protected by a kind of revetment on the sea or lake side to resist wave attack, and often with grass cover on crest and inner slope. Other countries may use terms like levees or embankments, but the structures are more or less similar.

The paper will cover past, present and future strategies. Interest in coastal flooding is increasing in the Netherlands and not only by coastal or civil engineers. Recently, this has widened the scope of feasibility studies to explore all kind of ideas, like insurance, evacuation, awareness, compartment (dividing a flood risk area in two parts, reducing the consequences of flooding) and also indestructible dikes. This last option would mean a flooding probability of (almost) zero and therefore a flood risk of almost zero.

As already noted, observations, conclusions, etc., made in this paper and the presentation are on the personal account of the author, so do not represent any official view from the Netherlands.

2 DECISIONS AFTER THE 1953 FLOOD

Early February, 1953, a severe storm hit the south west part of the Netherlands and also parts of Belgium and the UK, causing severe flooding with, in the Netherlands, almost 2000 casualties. Although people had warned before about the fairly low dikes and the real possibility of a major flood, the interest in those days after the world war was more directed to build up the country again, than on spending money for dike improvement.

After the flood the Delta Committee was formed with the main goal to present a safety policy against flooding for the future. In those days they performed a kind of flood risk analysis. They concluded that the probability of *flooding* for central Holland (Amsterdam, The Hague, Rotterdam) should be around 1/125,000 per year. But they wanted or had to be practical, and they understood that calculating probability of failure, including all failure mechanisms of dikes, was not yet possible.

The outcome was: design a *safe* dike for an *event* with a probability of 1/10,000 per year. This had two advantages. First it was clear for what kind of event the dikes should be designed and secondly, normal design procedures could be used (instead of describing failure mechanisms leading to flooding, which is necessary for flood risk design).

But the principal of flood risk was not forgotten. It was clear that some parts of the country had more inhabitants and more investments than other parts and consequences of flooding, therefore, would be different. Each part got his own "event" to design for: 1/10,000 per year for central Holland, 1/4,000 per year for most others and for smaller areas even 1/2,000 per year.

Later on, also the rivers were included in the safety policy. It was realized that evacuation would be possible for flooding from a high river discharge, as it would be predicted some days before. This would lead to less casualties and, therefore, most river dikes had to be designed for a water level which would have a probability to occur of 1/1,250 per year. Since then the safety against flooding always considers both, coastal flooding and river flooding. Figure 1 shows all primary flood defences.

It was Edelman (1954) who realized that if three weak points were present at a dike section under severe wave attack, it would fail:

1. If the crest was too low, this would lead to extensive overtopping;
2. If bad quality of material was present, infiltration of water in the dike would be fast;
3. If a steep inner slope was present, it would lead to a slip failure when wet.

Based on analysis of dike breaches in 1953, Edelman concluded that if one of these 3 items was not present, then very often there was no breach. His suggestion was to make inner slopes much more gentle, like 1:3, but allow overtopping. He was convinced that a dike could withstand wave overtopping, as long as the inner slope would be gentle enough.

The final decision for design, however, was different. It was indeed decided to make gentler inner slopes of 1:3, but moreover, not to allow (severe) wave overtopping. The crest height should be designed equal to the 2%-wave run-up level. It was expected that any dike crest and inner slope with grass cover would resist 2% of the incoming waves overtopping the crest.

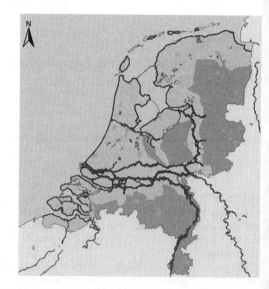

Figure 1. The Netherlands as delta with all primary flood defences, both for coastal and river protection.

With this design principle all sea dikes have been improved since 1953 and actually, present designs still use these principles. In the nineties the 2%-wave run-up level changed to 1 l/s per m wave overtopping.

3 SAFETY ASSESSMENT

After improvement of most of the dikes in the Netherlands and construction of the storm surge barriers in the Eastern Scheldt and the entrance to the port of Rotterdam, it was realized that the flood protection system should not only be designed and constructed, but should also regularly be checked. The Flood Defence Act of 1996 ruled that every 5 years a safety assessment should be performed on all primary flood defence assets.

This safety assessment has been based on the same principle as for the design: the dike or flood protection asset should be safe for a certain event with a certain probability of occurrence. But there are certainly differences between safety assessment and design.

In a design the actual properties of the material of the dike are not known, but assumed. Safety factors are taken into account and a little more safety does not cost a lot more as it will all be part of a new or improved structure. In a safety assessment the structure is present and material properties can be measured. But including more safety means that the present structure will be disqualified too early and will directly lead to large costs for improvement. The main principle of a safety assessment should be, also indicated in the Dutch safety assessment manuals, that: "A lot may go wrong, but the dike may not breach".

Reality is different. Experience shows that where doubt is present, the dike section will be disqualified. There is probably another reason behind these decisions, not stated publicly. The water boards have to maintain the majority of the dikes. They have to pay the maintenance from local taxes they earn. But major improvements, as a consequence of the safety assessment, will be paid by the government. It is for this reason that water boards can not completely be objective in the safety assessment procedure. There is benefit in obtaining an improved protection.

The safety assessment is quite complex as it has to consider all parts of a dike or flood defence asset, for all kind of failure mechanisms. Certainly in the first assessments, parts were discovered which did not pass the assessment criteria or where assessment criteria were not yet available. In the latter case also design rules were not available and actual design had always been based on experience rather than design rules.

When the results of the first assessments were summarized, it appeared that in about one-third of all the dike sections, parts were disqualified (and had or have to be improved) or an assessment rule was not available (and therefore no assessment result was available). This has been interpreted from two sides. One side states that even with disqualification of parts, there is not a direct threat for flooding and there will be sufficient time to design and improve the part of the dike, such as a stronger revetment or a little higher crest. Lacking knowledge means that this knowledge has to be developed. The other side states that only two-third of all flood defence assets are safe and that the other one-third gives a serious threat. So politicians should release more money for improving dikes and developing knowledge.

A more general conclusion is that the Netherlands has never been more safe against flooding than in the present situation, but that still quite some work has to be done to be safe in agreement with the safety assessment rules.

4 FROM PROBABILITY OF EVENT TO FLOODRISK

The present design and safety assessment rule is that the flood defence asset should withstand an event with a certain return probability or probability per year. It does not explicitly state that the defence asset should also withstand a (much) more severe event, it should only be designed with some or enough safety.

Of course this has led to a discussion on other possible normative rules. Since 1990 a discussion started on better safety codes: from probability of event (present situation) to probability of flooding (breaching of dikes) to flood risk. All three have their pro's and contra's. The probability of flooding is easy to explain to the public: it gives the probability per year that it is expected to get wet feet. The difficulty is that one needs a full description of each failure mode from initial damage up to the initiation of a breach.

Work under Task 4 of FloodSite (Allsop et al., 2007) made a good step by describing most of the failure modes. But the result of a calculation is never better than the failure mode modeled. VNK 1 was the first attempt to calculate flooding probabilities for various areas in the Netherlands. (VNK stands for Safety of the Netherlands calculated and mapped). Real dike ring areas were considered and all possible failure mechanisms. It took a few years by a number of consortia to come up with results for some 10–15 dike ring areas (where we have more than 50).

One conclusion or result was that by this procedure it is easy to find the weakest locations in a dike ring and for what failure mechanism. Upgrading that section directly reduces the probability of flooding. It might be noted, however, that these "weak" sections can also be found by applying the regular safety assessment. The probabilistic method, however, gives how much the probability of flooding would improve, which is not possible with the safety assessment.

The calculations by VNK 1 also showed that some failure mechanisms were not well understood or not modeled well enough. And in such a case uncertainty is taken into account which sometimes led to unrealistically large flooding probabilities. In such a case more study is required to improve the modeling of the failure mechanisms.

Since VNK 1 the modeling has improved and production runs will be made in 2008/2009 to calculate flooding probabilities of all 53 dike ring areas in the Netherlands under VNK 2.

During (and after) VNK 1, a lot more information became available on the consequences of flooding. Numerical tools were developed to model realistically the water flow and inundation in time, assuming one or more initial breaches in the dike ring system. Damages were calculated as well as casualties. Extreme assumptions were made to find upper boundaries. Moreover, it gave insight in inundation depths and the most vulnerable locations in the Netherlands.

Flood risk is the product of probability of flooding and consequences, so probability multiplied by cost (money). There is similarity with an insurance premium. A flood risk could be for example 2 million euro per year. It is not an easy definition to explain to public. It is also not easy to regulate flood risks in a normative rule. Although 15 years ago the final goal seemed to be to come to regulations based on true flood risk, nowadays the insight has changed a little. Probability of flooding, as calculated by VNK 2, will probably be taken as the primary result. In future flood probability may become the normative rule. The *insight* in consequences (damage, casualties) will steer the normative rule, not the product of probability and damage.

5 SAFETY UP TO 2100

Many feasibility studies are going on in the Netherlands and safety against flooding now has interest from a wider professional audience than just civil engineers. On 19 June 2008 a one day conference/workshop (general presentation, workshop discussions, no papers available) was held with the title "The power of water". This conference released the policy for flood defence in the Netherlands and consisted of 3 layers:

1. Prevention is and stays number one. It is always better to prevent anything to happen than to minimize the consequences. More knowledge should be gained on the actual strength of flood defence assets, consequences should be studied and climate change should be taken into account. Innovative solutions should be studied, like indestructible dikes.
2. Spatial planning should include safety against flooding.
3. Reduce remaining consequences by evacuation and awareness.

All points will be elaborated a little more, starting with the new points 2 and 3. Policy makers believe that spatial planning can work if a safety assessment procedure will be part of it. It should lead to decisions not to built new houses or industry in some parts, where for example the area is many meters below sea level. Or it should lead to decisions to raise the level several meters before starting construction.

A large part of the conference did not believe that this second layer would work. The main reason is that spatial planning is in the hands of the local authorities who decide on it, not the government. Local authorities will always decide to improve their own area and will never say: go to the town 10 km further, because their level is higher than here! Another reason is that flooding by sea or river is not an issue in daily politics of a local area.

Evacuation belongs to the third layer. Due to the fact that discussion on flooding always includes both sea and river flooding, some interpretations of phenomena are considered true in both situations. For evacuation this is certainly not the case and only a few people are aware of it.

A high river discharge in the Netherlands, with consequently a high water level against the dikes, is not caused by flash floods, but by very heavy rain in Switzerland and the south of Germany, and probably the Netherlands. It takes days before this water comes to the border of the Netherlands and good computer models are available to predict where, when and how high the water level will come along the river dikes.

In case of an emergency, where predicted water levels may indicate an unsafe situation, there is time to evacuate thousands of people. In 1995, 100,000 people were evacuated in a situation where some dikes were not yet improved and where the safety could not be guaranteed during that high water. In those cases the weather is not too bad for evacuation and there is time enough.

Also for a hurricane, like in the US, there is time to evacuate. Evacuation in such a situation, however, is mainly based on the destructive wind along the coast, not entirely on a probability of flooding.

The possibility of evacuation is often transferred to coastal situations. And this is a complete mistake, certainly for Dutch situations! It may be possible in the UK in rural areas, where for example a small number of farmers live in a relatively small flood risk area, protected by dikes only able to protect against events smaller than 1/30 or 1/100 years (or even less). For each very severe storm warning they should evacuate. But here it will be a very small number of people who are aware of the situation.

Assuming that the dikes in the Netherlands can withstand an event with a return period of 10,000 years, evacuation would only be an option if a storm is expected which would even be worse. At present we are not able to predict whether a storm would be an event with a smaller or larger return period than 10,000 years. It depends also on the local conditions like tide where the worst condition with respect to maximum surge level and wave conditions will occur. This means that we should evacuate the whole north, west and south west of the Netherlands, say around 5–10 million people, for a storm warning with a return period in the order of 1,000 years or more.

But would that be possible? Such a severe storm will already have a wind force close to Beaufort 11 one day before the actual peak of the storm (with then certainly Beaufort 12 and more). Such a 1/10,000 years storm will have a devastating effect on the country. Many roofs will blow away, thousands of trees will break, tiles and everything which was not tied thoroughly, will fly through the air.

In 1999 a short but very strong storm hit the Atlantic coast north of Bordeaux in France. Large areas with trees were completely destroyed. Even after more than two years the trees were not yet all removed, see Figure 2. The storm led to flooding in the Gironde.

Figure 2. Two and a half years after a short but devastating storm, all fallen trees have not yet been removed. Gironde, France, June 2002.

It may be clear: nobody wants to evacuate in such a storm. It would be very dangerous. The only option is to wait in a safe place and, indeed, if a flooding occurs, go to the first or second floor and hope that the house will be strong enough to withstand the water.

So in planning any such evacuation, the division between coastal flooding and river flooding must be made, and it may be wiser not to assume that evacuation is always possible.

It is, however, always good to increase awareness for disasters, like a flooding, which is the second item of the third layer (evacuation and awareness). During such an event power may be shut down, as well as energy, water supply, etc. Awareness and preparations are good, not only for a disaster like flooding, but actually for all possible disasters.

6 SAFETY OF COASTAL DIKES

6.1 Main failure mechanisms

Coastal dikes are designed for high storm surges and related severe wave attack. Both the high water level and the waves give the loading to the dike. Two main failure modes exist for coastal dikes. One is the failure of the seaward protection by large waves. A many small and large scale model tests have been performed in wave flumes to find the relationship between wave attack and strength of a variety of revetments, from rock revetments to asphalt layers. We can conclude that we know a lot on strength of these kind of protection systems.

The other main failure mechanism is wave overtopping and failure of the inner slope of the dike. We know a lot about wave overtopping, or actually, the hydraulic behaviour of waves overtopping a dikes, see the Overtopping Manual, 2007. But we have only little experience about how strong dikes are against wave overtopping. This is simply because small scale model testing is not possible, due to the fact that clay and grass can not be scaled down. Until recently, only large scale testing in the Delta flume (the Netherlands) or GWK (Germany) have been options and indeed some tests have been performed in these facilities, in the past and recently.

The fact that the hydraulic behaviour of wave overtopping is known, has led to the idea of the Wave Overtopping Simulator. This new device has been used for erosion tests performed on several real dikes and insight in strength has gained tremendously. Results will be summarized here.

6.2 Erosion by wave overtopping

Two mechanisms may lead to failure due to wave overtopping. The first is infiltration of overtopping water into the dike and eventually sliding of the inner slope. The second is erosion of the cover layer of clay and grass by overtopping waves, followed by erosion of the inner slope (clay or clay layer on sand core).

The first mechanism, infiltration and sliding, can only occur if the inner slope is quite steep, see also the points mentioned by Edelman, 1954, in Chapter 2. For this reason most coastal dike designs in the Netherlands, after the flood of 1953, got a 1:3 inner slope. It is assumed that such a slope will not slide due to infiltration of water. But if a steeper slope is present, already 1 l/s per m overtopping would be enough to give sufficient infiltration of water.

This means that for steep inner slopes (steeper than 1:3 or may be 1:2.5) the critical overtopping discharge is already 1 l/s per m. For dikes with an inner slope of 1:3 or gentler we assume that infiltration and sliding is not a governing failure mechanism. Only erosion by overtopping remains.

Till a few years ago hardly anything was known about resistance of inner slopes of dikes with grass against wave overtopping. But in the beginning of 2007 and 2008 innovative erosion tests have been performed for various dike sections. In 2006 the Wave Overtopping Simulator was constructed, see Van der Meer et al., 2006. The basic idea is that a constant discharge is pumped into a box on top of a dike and then the pumped volume is released from time to time in such a way that it simulates overtopping waves in reality. Figure 3 gives an impression of the working of this wave overtopping simulator.

Tests have been performed for mean overtopping discharges starting at 0.1 l/s per m up to 75 l/s per m. In 2007 3 dike sections have been tested, which are reported by Van der Meer et al., 2007, Akkerman et al., 2007 and in the ComCoast reports (*www.comcoast.org*). In early 2008 another 9 dike sections were tested (see Figure 4) at three locations in the Netherlands.

Figure 3. The Wave Overtopping Simulator releases 22 m³ of water over 4 m width in about 5 s. It simulates a large overtopping wave with a mean discharge of 75 l/s per m.

Figure 4. Damage to a dike section during a test with 75 l/s per m wave overtopping.

Part of the results has been given by Steendam et al., 2008, at this conference. They come to a few preliminary conclusions, mainly based on observation rather than thorough analysis, which still has to be performed. The most important one in relation to actual strength of dikes by wave overtopping is:

It seems unlikely that an inner slope with a clay cover topped with a grass cover (in Dutch situations) will fail due to erosion by overtopping waves with a mean discharge of 30 l/s per m or less. Future research may result in a final conclusion.

A large number of dike sections withstood 50 l/s per m and some of them even 75 l/s per m. No section failed for 30 l/s per m, which gives the basis for the preliminary conclusion.

7 INDESTRUCTIBLE DIKES

7.1 Case study

The 10^{-4} event is already very extreme. In stochastic terms a probability of zero does not exist, but "practically zero" can be defined as: two orders of magnitude more safe than now. If a dike can resist a 1/1,000,000 storm can we give it the title indestructible? What do we have to do to make such a dike?

A short feasibility study was made to explore this idea. Four cases (dike sections) were chosen, one in the north along the Waddensea, one directly on the North Sea coast, one in an estuary and one along the coast of the big lakes. All cases showed for the safety assessment situation (event around 1/10,000 per year) an overtopping discharge around 1 l/s per m.

Wave conditions and water levels were determined for the 10^{-4}, 10^{-5} and 10^{-6}-events and then PC-OVERTOPPING was used to calculate the overtopping discharges. These were respectively around 1, 5–10 and 20–30 l/s per m. The 20–30 l/s per m overtopping discharge is still equal to or smaller than the limit of 30 l/s perm.

A preliminary conclusion may be that a design with 1 l/s per m overtopping leads to a robust and "indestructible" dike section (with respect to erosion by overtopping). It should be noted that such a dike should have an inner slope of 1:3 or gentler.

A more extreme event does not only lead to higher water levels, but also to larger waves. Another failure mechanism is stability of the revetment. Most stability formulae are based on the stability number $H_s/\Delta D$, where H_s = the significant wave height (at the toe of the dike), Δ = relative mass density and D = a diameter or thickness.

A larger wave height leads then linearly to a larger diameter or thickness. The increase in wave height from a 10^{-4} to a 10^{-6}-event is more or less the same increase that is required to make the revetment "indestructible". In the case study the increase in wave height was 10–25%. The consequence to make an "indestructible" revetment would be to increase the thickness by at least 10–25% and also to apply the revetment protection to a higher level on the dike, as the 10^{-6}-event has a higher water level.

The conclusion might be that if coastal dikes can already resist a 10^{-4} storm, indestructible dikes are may be closer to become reality than we thought. Moreover, it is already tradition during the past 300 years that every one or two generations the dikes have been improved. There is no reason to believe that this tradition will stop. Improvements in the past few decades have always been designed for a life time of 50 years. It can be assumed that in the next 50 years almost all coastal dikes, or at least a majority, in the Netherlands will be improved again. That is a unique opportunity to investigate and go for indestructible dikes.

It is realized that this is perhaps a situation which is only present in the Netherlands. It is different in situations where the present safety is 1/100 per year or less. But even there, prevention is always better than facing a major flood.

7.2 *Fragility curves*

Safety assessments of flood defence assets are increasingly performed with the technique of structural reliability. All parameters, load parameters (hydraulic boundary conditions) and strength parameters (dike characteristics), are taken into account and expressed as stochastic variables. One of these structural reliability methods is to calculate the failure probability (P_f) of a flood defence, *given a certain water level*. Assembling the failure probabilities for several water levels constructs a fragility curve, see Van der Meer et al., 2008, presented at this conference.

This paper described the situation in the Netherlands, where design events have a return period in the order of 10^{-4} per year. The fragility curve gives the probability of failure given a certain water level, not a return period of that water level. But in an actual case there is a known relationship between the water level (storm surge), including wave conditions, and the return period of that event. Therefore, it is fairly easy to calculate a fragility curve where the probability of failure is give as a function of the return period of the water level or event. Figure 5 gives an example for a large sea dike (one of the case studies discussed before).

The graph shows actually three failure modes:

1. Infiltration of overtopping water and sliding of the inner slope (if the inner slope would be steep). This would occur for an overtopping discharge of 1 l/s per m;
2. Erosion of the inner slope by wave overtopping (the curves with overtopping discharges of 10–50 l/s per m);
3. Piping.

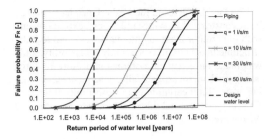

Figure 5. Fragility curves as a function of the return period of the water level.

Piping in this example does not give a serious probability of failure. The 1 l/s per m overtopping discharge gives more or less a probability of failure of 50% for the 10^{-4}-event. This is exactly the design condition.

But the graph gives also a similar impression as the calculations on indestructible dikes: the 50%-probability for 30 l/s per m in this graph gave a return period of 2.10^{-6}, which is more extreme than the 10^{-6}-event. One can say that the differences between the curves for 1 and 30 l/s per m in Figure 5 give the safety between design and failure and that the probabilities for the 30 l/s per m curve actually indicate that this dike section is "indestructible" with respect to erosion by wave overtopping.

8 CONCLUDING REMARKS

The major improvements of coastal dikes in the Netherlands, after the 1953 flood, was based on three principles. Design for an event with a return period around 10,000 years; make inner slopes of a dike at least 1:3; and design for the 2%-run-up level or 1 l/s per m wave overtopping. This has led to high and strong dikes.

A safety assessment procedure was introduced, which has to be performed every 5 years for all flood defence assets. The first assessments showed weak and inadequate parts, which are still being improved.

A new policy on flood defence was released recently, where three layers were introduced. The first still being prevention. The two added layers are to include safety against flooding in spatial planning and to make evacuation plans and to make people more aware of the possibility of a disaster. These two added layers still have to be explored.

The recent destructive tests with the Wave Overtopping Simulator showed that clay with a grass cover on the inner slope of a dike is well resistant to wave overtopping. More resistant than many people thought, including the author.

The fact that dikes in the Netherlands have already been constructed to withstand a very extreme event, with only minor overtopping, makes the step to indestructible dikes within reach. Only a feasibility study was made on a few case studies and more research is required to investigate all consequences, including costs. But it certainly is an opportunity if the next 50 years most coastal dikes will have to be improved once more.

ACKNOWLEDGEMENTS

The Directorate-General Water of the Rijkswater-staat and Deltares are acknowledged for the possibility to explore the idea of indestructible dikes. The Centre for Water Management of the Rijkswaterstaat and the project group on Erosion by Wave Overtopping, including Deltares, Infram, Royal Haskoning and Alterra are greatly acknowledged for recent overtopping tests at the Wadden Sea and the cooperation to come to early and preliminary conclusions on the test results. ComCoast is acknowledged for their support to develop the Wave Overtopping Simulator and to perform the first tests in Groningen. Finally, the Project Organization of Sea Defences (Projectbureau Zeeweringen) is acknowledged for their support to perform the overtopping tests in the south west of the Netherlands.

REFERENCES

Akkerman, G.J., P. Bernardini, J.W. van der Meer, H. Verheij and A. van Hoven, 2007. *Field tests on sea defences subject to wave overtopping.* Proc. Coastal Structures, Venice.

Allsop N.W.H., A. Kortenhaus and M. Morris, 2007, *Failure Mechanisms for Flood Defence Structures,* Task 4 report (nr.T04-06-01) Wallingford: FloodSite

Edelman, T., 1954. *Doorbraakvrije zeedijken (in Dutch, translated: Indestructible sea dikes).* Note 1954.

Overtopping Manual, 2007. Pullen T., N.W.H. Allsop, T. Bruce, A. Kortenhaus, H. Schüttrumpf and J.W. van der Meer. *EurOtop: Wave Overtopping of Sea Defences and Related Structures: Assessment Manual* (pdf download available from: *www.overtopping-manual.com*)

Steendam, G.J., W. de Vries, J.W. van der Meer, A. van Hoven, G. de Raat and J.Y. Frissel, 2008. *Influence of management and maintenance on erosive impact of wave overtopping on grass covered slopes of dikes.* Proc. FloodRisk, Oxford, UK.

Van der Meer, J.W., W.L.A. ter Horst and E. van Velzen, 2008. *Calculation of fragility curves for flood defence assets.* Proc. FloodRisk, Oxford, UK.

Van der Meer, J.W., P. Bernardini, G.J. Steendam, G.J. Akkerman and G.J.C.M. Hoffmans, 2007. *The wave overtopping simulator in action.* Proc. Coastal Structures, Venice

Van der Meer, J.W., P. Bernardini, W. Snijders and E. Regeling, 2006. *The wave overtopping simulator.* ASCE, proc. ICCE, San Diego.

TECHNICAL PRESENTATIONS

Inundation modelling

Recent development and application of a rapid flood spreading method

Julien Lhomme, Paul Sayers, Ben Gouldby, Paul Samuels & Martin Wills
HR Wallingford Ltd., Wallingford, Oxfordshire, UK

Jonatan Mulet-Marti
Wallingford Software, Wallingford, Oxfordshire, UK (formerly HR Wallingford)

Flood risk analysis involves the integration of a full range of loading, multiple defence system states and uncertainty related to the input parameters of the model. This type of analysis involves the simulation of many thousands of flood events. To keep model runtimes to practical levels an efficient yet robust flood inundation model is required.

This paper describes recent improvements to a rapid flood spreading model. The model is based on a storage cell concept, defined as Impact Zones (IZ) that are based on topographic depressions in the floodplain. The Impact Zones are built in a pre-processing step using the DTM of the considered area. The pre-processing also calculates the IZ characteristics: (i) relation between the water level and the volume stored in each IZ, (ii) relations and Communication Levels between neighbour IZ. The Impact Zones are then used as the elementary units for the spreading of the water.

Any given flooding scenario is defined by specifying Input Volumes at the boundary IZ. These can be defined by the user or can be calculated by a breach/overtopping module. At the start of the spreading process, the IZ with excess volume are identified by comparing the IZ capacity and its input volume. Then any Excess Volume is spilled from the concerned IZ into one or more of its neighbours. Two or more neighbour Impact Zones having the same Water Level are merged into a single IZ. This Spilling/Merging process is repeated until there is no more excess volume in any IZ. The computed flood extent is considered as the final state of the flood.

Recent developments to the model include a better representation of multiple spilling IZs, to improve the flood extent, and a method to improve the representation of the route that floodwater takes over the floodplain area, known as the flood "pathway". This improved model is applied to a number of different sites with comparisons made to more complex models and to observed data. The findings of this comparison demonstrate a good degree of similarity between the RFSM and more complex models, with a significantly reduced runtime overhead.

Keywords: Flood risk assessment, flood spreading, computational time, storage cells model

Hydrodynamic modelling and risk analysis in RAMFLOOD project

Ernest Bladé, Manuel Gómez-Valentín, Josep Dolz & Martí Sánchez-Juny
*FLUMEN Research Group, E.T.S. d'Enginyers de Camins, Canals i Ports de Barcelona,
Universitat Politècnica de Catalunya, Barcelona, Spain*

Javier Piazzese, Eugenio Oñate & Georgina Corestein
*CIMNE International Center for Numerical Methods in Engineering, Universitat Politècnica de Catalunya,
Barcelona, Spain*

The aim of the RAMFLOOD project (IST Programme of the European Commission, 5th Framework Programme) was to construct a decision support system (DSS) for the risk assessment and management of emergency scenarios to assist public administrators and emergency services. A crucial component of the system was the hydrodynamic simulation module which was the base for the following hazard assessment as well as the training of an artificial intelligence module.

For this hydrodynamic modelling CARPA was used. CARPA is a hydrodynamic simulation system developed by Flumen research group of UPC. The system solves the full Saint-Venant equations in one and two dimensions using a shock capturing explicit high resolution finite volume method. The numerical scheme in CARPA is based in the WAF TVD scheme, which can be also seen as an extension to systems of equations of the Lax-Wendroff scheme, or also as a second order extension of Roe scheme. Unstructured meshes of quadrilateral and triangular elements can be used for the 2D solution, while in the 1D domain the classical cross section discretisation is used. Mixed 1D-2D simulation allows for more efficient numerical schemes, using low computing cost where 1D flow is assumed, combined with the precision of a high resolution 2D scheme in floodplains or even main channel if flow patterns are indeed two-dimensional.

The integration of 1D and 2D schemes is achieved straightforwardly thanks to the explicit schemes used. The technique to impose boundary conditions to either domain that has been used consists in the common idea of assuming a set of fictitious elements or boundary finite volumes just outside the domain boundaries. Values of the required hydraulic variables are given in such boundary elements.

In 1D-2D connections there is an overlapping of boundary 1D and 2D elements. In such way, when computing one time step the values of water elevation and velocity at the 2D boundary elements are obtained from the last (or first) section of a 1D reach at the previous time step, while the values of discharge and cross section area in the boundary 1D elements are obtained from the addition of the 2D elements corresponding with that section position.

Thanks to the integrated 1D-2D calculation time reduction is achieved, which is of capital importance in the DSS training system as a great number of simulations are needed.

Risk criteria considered is associated either to human risks in terms of critical water levels, velocities or combination of both parameters, and to time of water permanence. According to these criteria, a risk database was developed. For the first one the following three different risk levels were considered: High Risk, Moderate Risk and No Risk according to the Catalan Water Agency (ACA) criteria. The system was applied to a 30 km long reach of Llobregat river near Barcelona.

Keywords: Flood risk management, hydrodynamic simulation, 1D-2D integration

Testing and application of a practical new 2D hydrodynamic model

J. Gutierrez Andres, J. Lhomme, A. Weisgerber, A. Cooper & B. Gouldby
HR Wallingford Ltd., Wallingford, Oxfordshire, UK

J. Mulet-Marti
Wallingford Software, Wallingford, Oxfordshire, UK

Making Space for Water is a cross Government programme to take forward the developing strategy for flood and coastal erosion risk management in England. As part of this programme, a significant amount of effort is being directed towards finding ways to improve the management of urban drainage to reduce flooding. This need was brought sharply into focus during the summer floods of 2007, when the City of Hull suffered from severe flooding due to an overwhelmed drainage system. It is recognized that an integrated approach is required, linking rivers and their floodplains with surface water and foul drainage and as a result of this modelling methods are starting to converge.

River modellers are increasingly modelling fluvial flood events with linked 1D (river channel) and 2D (floodplain) hydrodynamic models.

Drainage modellers have long understood that accurate modelling of extreme urban flood events requires a better understanding of overland flow paths and the capability of representing the re-entry of surface flows into the below ground drainage network. But due to the complexity of both the underground and above ground systems within urban areas, up to now there has not been an easy way to represent both systems interactively.

InfoWorks2D has recently been released by Wallingford Software. This 2D hydrodynamic modelling software allows linkages with the already existing 1D software for rivers (InfoWorksRS) and network systems (InfoWorksCS). The main characteristics of the 2D component are:

a. Finite volume formulation (weak solution of the shallow water equation).
b. Based on the Gudonov scheme and the Riemann solvers (Shock capturing scheme).
c. Use of an unstructured mesh.
d. Fully integrated with the 1D existing engine.

The 1D components have been in use, worldwide, for many years and are well known and well tested. This paper describes the results of a series of analytical tests used to validate the robustness of the new 2D modelling engine, results of tests against observed flood data and, in some cases, comparisons of the results with other 2D flood spreading models. This paper also discusses the potential benefits of applying integrated 1D/2D modelling techniques to the analysis of flood events.

Keywords: 2D hydrodynamic model, fluvial flooding, integrated urban drainage, InfoWorks

Floods study through coupled numerical modeling of 2D surface and sewage network flows

C. Coulet, L. Evaux & A. Rebaï
SOGREAH Consultants, Echirolles, France

The characterisation of flood risks in urban areas stems from the combination of relevant hydrological scenarios and hydraulic model simulations of the study area. The quality of the results depends on both the hypotheses adopted and the representativeness of the models used. In this context the RIVES research project, backed by the ANR (the French national research agency), is aimed at identifying the key parameters to be taken into account when studying urban flooding, producing representative hydrological scenarios, and improving the modelling resources used in this field.

The sewerage network, which discharges stormwater and can cause flooding as a result of water backflow, is therefore being considered as part of the third research area of the RIVES project. Sogreah worked on this particular issue as part of the project, setting up dynamic coupling between the TELEMAC-2D surface flow numerical modelling system and the CANOE sewerage network modelling system.

The work completed is based on a detailed bibliographical study of flows between the surface and the network in order to obtain a simplified discharge capacity formulation for central and lateral drainage outlets (weir or orifice type, streaming flow or free flow, pressurised or unconfined). It is now possible to specify the dimensions of each drainage outlet, the discharge coefficient and any obstruction of the flow area.

The coupling is based on existing functions in the two computation codes, through exchange points: "overflow points" in CANOE and "source points" in TELEMAC-2D. In practice, the coupling is carried out using a fractional step method:

- A first step corresponding to the network flow during which CANOE calculates an exchange discharge (at time t + dt) on the basis of water depth information supplied by TELEMAC-2D (at time step t);
- A second step corresponding to the surface flow during which TELEMAC-2D takes account of the resulting (positive or negative) exchange discharges from CANOE (at time t + dt) to calculate the water levels (at time t + dt).

If this method is to be made more representative, it should be modified as part of future coupling upgrades in order to implement a temporal alternation (resolution offset by dt/2 between the surface module and the module taking account of the pipe network).

First of all the coupling was validated on simplified test cases, then on an actual case study for a district of Oullins, a town on the river Yzeron in the Rhône *département* of France. This validation made it possible to check the robustness of the coupling and its sensitivity to certain numerical parameters: mesh size, simulation time step, coupling time step, etc.

The results obtained must now undergo further tests on actual case studies for which more comparison data is available, in terms of both surface flows and network flows. However, the developments achieved prove that the coupling of systems used to model 2D surface flows and flows in stormwater drainage networks is practically operational, and it should be possible to incorporate this quickly and easily into a commercial version of the software applications in the near future.

Keywords: Runoff, flood, stormwater risk, sewerage, hydroinformatics, numerical modelling, coupling, Telemac, Canoe

Modelling of flooding and analysis of pluvial flood risk – demo case of UK catchment

J.P. Leitão, S. Boonya-aroonnet & Č. Maksimović
Department of Civil and Environmental Engineering, Imperial College, London, UK

R. Allitt
Richard Allitt Associates Ltd., Haywards Heath, West Sussex, UK

D. Prodanović
Faculty of Civil Engineering, University of Belgrade, Serbia

Insufficient capacity of storm drainage system (especially during extreme events) is commonly one of the causes of urban flooding. In many cases, exceeding (surcharged sewers) flow returns from the sewer system to the surface (major system) and interacts with the excess surface runoff.

Understanding of vulnerable locations and flood paths in urban areas has been enabled by physically based modelling of surface runoff. It relies on detailed analysis of preferential flow patterns influenced by land use, built structures and terrain's elevation data (DTM). This type of analysis helps planners or designers to conceptualise better and safer drainage systems for the future and city managers to minimise damages caused by urban pluvial flooding.

This paper describes concepts and implementation of a GIS-based tool for modelling of overland flow and urban flood which has been developed at the UWRG (Urban Water Research Group), Imperial College London and at the Faculty of Civil Engineering, University of Belgrade. The tool can be coupled with existing models for urban storm drainage analysis including "UK industrial standards" for this example InfoWorks CS. In addition to presenting the developed methodology the paper presents the results of its application in a real case study in the Cowes catchment, Isle of Wight, UK. In this urban area, flooding events have been reported over the past decades. Numbers of historical records are available and have been used in model refinement, calibration and demo applications.

The paper deals with the following:

- creation of the overland flow 1D network for modelling of urban pluvial flooding integrated with 1D pipe flow,
- test of developed GIS-based tool applicability, and
- comparison of the results achieved using the surface flow network generated by the GIS-based tool with the common procedure.

The methodology is based on the detailed analysis of Digital Elevation Model (DEM) in order to define the surface flow network. The surface flow network consists of temporary ponds and connecting pathways which are integrated in the integrated pluvial flooding simulation tool. The key parts of the methodology can be identified as follow:

- enhancement of DEM for path generation processes,
- delineation of local depressions,
- delineation of flow pathways,
- estimation of pathways' geometry, and
- generation of input files for simulation of interactions with surcharged sewer network.

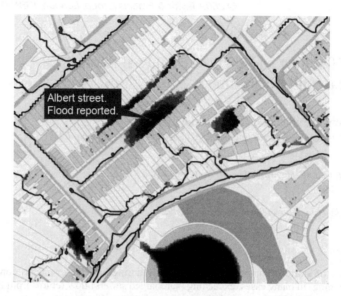

Figure 1. An example of surface flow network created by the developed methodology.

Figure 1 illustrates an example in which numbers of surface depressions are identified in the area and the tool could locate a depression at the Albert St. where flood events have been recorded historically.

The work carried out so far and presented in this paper shows the applicability of developed methodology. It allows the city planners and the urban drainage modellers to manage urban pluvial flood risks more realistically and efficiently.

Keywords: Flood risk management, GIS, Pluvial Urban Flooding, Surface Flow Network

An integrated approach to modelling surface water flood risk in urban areas

J.B. Butler, D.M. Martin & E.M. Stephens
Ambiental Technical Solutions Ltd., Brighton, East Sussex, UK

L. Smith
University of Sussex, Brighton, East Sussex, UK

Two thirds of the flooding that occurred in the UK during summer 2007 was as a result of surface water rather than river or coastal flooding. In response, the Environment Agency and Interim Pitt Reviews have highlighted the need for surface water risk mapping and warning tools to identify, and prepare for, flooding induced by heavy rainfall events. This need is compounded by the likely increase in rainfall intensities due to climate change. The Association of British Insurers has called for the Environment Agency to commission nationwide flood risk maps showing the relative risk of flooding from all sources. At the wider European scale, the recently-published EC Directive on the assessment and management of flood risks will require Member States to evaluate, map and model flood risk from a variety of sources. As such, there is now a clear and immediate requirement for the development of techniques for assessing and managing surface water flood risk across large areas.

This paper describes an approach for integrating rainfall, drainage network and high-resolution topographic data using Flowroute™, a high-resolution flood mapping and modelling platform, to produce deterministic surface water flood risk maps. Information is provided from several European case studies to enable assessment and validation of modelled results.

Flowroute was co-developed with flood scientists at Cambridge University specifically to simulate river dynamics and floodplain inundation in complex, congested urban areas in a highly computationally efficient manner. It utilises high-resolution topographic information to route flows around individual buildings so as to enable the prediction of flood depths, extents, durations and velocities. As such, the model forms an ideal platform for the development of surface water flood risk modelling and mapping capabilities.

The 2-dimensional component of Flowroute employs uniform flow formulae (Manning's Equation) to direct flow over the model domain, sourcing water from the channel or sea so as to provide a detailed representation of river and coastal flood risk. The initial development step was to include spatially-distributed rainfall as a new source term within the model domain. This required optimisation to improve computational efficiency, given the ubiquity of 'wet' cells early on in the simulation.

Collaboration with Thames Water Utilities has provided detailed drainage information, and from this a simplified representation of the drainage system has been included in the model via the inclusion of sinks and sources of water from the drainage network. This approach has clear advantages relative to a fully coupled method both in terms of reduced input data requirements and computational overhead.

The Flowroute surface-water flood risk platform enables efficient mapping of areas sensitive to flooding from high-intensity rainfall events due to topography and drainage infrastructure As such the technology has widespread potential for use as a risk mapping tool by the UK Environment Agency European Member States water authorities local governments and the insurance industry.

Keywords: Surface water flooding, Flood inundation modelling, Risk mapping

Estimation of flood inundation probabilities using global hazard indexes based on hydrodynamic variables

G.T. Aronica
Dipartimento di Ingegneria Civile, Università di Messina, Messina, Italy

P. Fabio, A. Candela & M. Santoro
Dipartimento di Ingegneria Idraulica e Applicazioni Ambientali, Università di Palermo, Palermo, Italy

Identification, assessment and mitigation of flood risk require a rich fund of knowledge and should be based on a thorough uncertainty analysis. Risk-oriented method is the most complete approach that can be used in the fields of flood design and flood risk management and it allows us to evaluate the cost-effectiveness of prevention measures to optimise investments. The usual procedure is to deal with a precipitation-runoff model and to associate to risk the same return period of original rainfall, in accordance with the iso-frequency criterion. Alternatively a flood frequency analysis to a given record of discharge data is applied, but again the same probability is associated to flood discharges and respective risk. This apparently simple approach has a number of pitfalls and uncertainties, due to fact that only the flood discharges cannot give a reliable evaluation of hazard, depending also by the global characteristics of the flood event. To overcome these problems the application of a new Monte Carlo simulation procedure to deriver flood hazard maps is here presented.

The layout of the procedure can be resumed as follows:

1. stochastic input of flood hydrograph modelled through a direct Monte-Carlo Simulation based on flood recorded data. This choice has been preferred to circumvent the uncertainties related to precipitation-runoff modelling. Generation of flood peaks and flow volumes has been obtained via copulas, which describe and model the correlation between these two variables independently of the marginal laws involved. The shape of hydrograph has been generated on the basis of a historical significant flood event;
2. modelling of flood propagation using a hyperbolic model based on the DSV equations (Aronica et al., 1998). The conservative mass and momentum equations for two-dimensional shallow-water flow, when the convective inertial terms are neglected, has been solved using a finite element technique with triangular elements;
3. definition of global hazard indexes based on hydrodynamic variables (i.e., flood water depth and flow velocities)

The procedure was tested on a flood prone area located in the southern part of Sicily, Italy.

Keywords: Flood risk, flood propagation, Monte Carlo simulation

Flood modeling for risk evaluation – a MIKE FLOOD vs. SOBEK 1D2D benchmark study

P. Vanderkimpen
Flanders Hydraulics Research, Flemish Government, Antwerp, Belgium
Soresma, Antwerp, Belgium

E. Melger
Deltares/Delft Hydraulics, Delft, The Netherlands

P. Peeters
Flanders Hydraulics Research, Flemish Government, Antwerp, Belgium

1 OBJECTIVE

Floods resulting from the failure of a water defence can be described by a combination of 1D and 2D modelling techniques. A 1D approach is used to describe breach growth and breach flow and a 2D approach is used to predict flood propagation in the inundated areas. As breach development may be influenced by backwater effects, breach flow and flood propagation should be evaluated simultaneously.

The flood modelling packages MIKE FLOOD (DHI) and SOBEK 1D2D (Delft Hydraulics) both offer the possibility to dynamically link a 1D breach model (MIKE11 or SOBEK 1D) to a 2D flood plain model (MIKE21 or SOBEK 2D). Both packages were compared to each other as part of an effort to estimate the contribution of model uncertainty to overall flood risk uncertainty.

2 1D BREACH GROWTH

SOBEK 1D and MIKE11 allow specification of breach growth (depth and width) as a function of time. A major drawback of user-defined time series is that breach growth may not be consistent with prevailing boundary conditions: a breach can continue to grow even if there is no breach flow.

SOBEK offers the possibility to simulate breach growth by means of the Verheij-vanderKnaap equation, an empirical breach growth equation which was established and validated after a number of laboratory and field tests. In MIKE11 an erosion-based breach growth model based on the Engelund-Hansen sediment transport equation is available.

3 2D FLOOD PROPAGATION

SOBEK 2D and MIKE21 allow the use of a rectangular DEM, which can be given land-use related roughness values. MIKE21 allows the user to specify an eddy viscosity whereas the current version of SOBEK 2D does not take eddy viscosity into account. Both models are able to account for wind friction.

4 1D-2D MODEL LINKING

In SOBEK, 1D and 2D model components are linked by means of an implicit numeral scheme. In MIKE FLOOD, on the other hand, the links are explicit and a user-imposed time delay (smoothing factor) may be necessary to ensure numerical stability.

5 SOFTWARE COMPARISON

When using time series to describe breach growth, breach flow hydrographs produced by both modelling packages turn out to be very similar.

Although both breach growth models differ significantly, MIKE11's erosion based breach growth model could be calibrated in such a way that it produces results comparable to those obtained with the Verhejj-vanderKnaap equation.

When identical or equivalent data (DEM, surface roughness) are used and physical process parameters (eddy viscosity, wind friction) are matched as closely as possible, both flood propagation models produce similar results (flooded area, maximum flood depth, maximum flow velocity).

6 CONCLUSION

The contribution of model uncertainty or parameter uncertainty to overall flood risk uncertainty can be considerable, but the "software uncertainty" resulting from the choice of a software package in which the selected models and parameters are implemented becomes insignificant when both packages provide similar possibilities. As a result, it can be stated the choice of the modelling package is in no-way predominant in flood risk evaluation.

Comparing forecast skill of inundation models of differing complexity: The case of Upton upon Severn

Kondapalli Srinivas
Inland Waterways Authority of India, Noida, U.P., India

Micha Werner
Department of Water Science & Engineering, UNESCO-IHE, Delft, The Netherlands
Deltares-Delft Hydraulics, Delft, The Netherlands

Nigel Wright
Department of Water Science & Engineering, UNESCO-IHE, Delft, The Netherlands

Inundation models have been recognized as useful tools for indicating the extent of flooding due to for example a design flood event. Despite this, inundation models have till date only been applied rarely within the operational forecasting environment. One reason for this is the extensive computational resources required, but as their development continues, it is clear that these constraints are quickly becoming less relevant. There are also various approaches available for simulation of inundation extent, ranging from simple projection of the results of a one-dimensional hydrodynamic model in the 2D plane, through simple 2D approaches available in codes such as LISFLOOD-FP, through to more complex models that allow simulation of the inundation event in the floodplain using a full hydrodynamic 2D approach, while modelling the main channel flow in one dimension. This paper investigates the reliability of inundation models of differing complexity in the forecasting environment. Three model approaches of different complexity are applied to a reach of the Severn River at the town of Upton upon Severn. These are calibrated using observed flood extent data from the November 2000 event. The models are subsequently applied with forecast hydrographs as inputs. These hydrographs are obtained from the current operational forecasting system for the river reach under consideration, with forecasts being made at differing lead times. Results show that calibration of the differing model types using the observed data results in reasonable performance for all three. There are some indications the integrated 1D-2D model describes complex inundation and subsequent drainage patterns around Upton upon Severn most adequately. When running within the forecast environment, this changes, and the performance of the different models diminishes due to uncertainties in the flow forecasts. The results from the three models becomes more comparable in the forecast environment, as the uncertainties in the inflow hydrograph start to dominate. Interestingly there are still significant differences in the recession after the flood-peak. A reliable description of this recession may be a requirement of for example the emergency services, showing that the model approach most suited is tightly knit with the use results from that model in the flood warning process.

Keywords: Flood forecasting, inundation modelling, model complexity

Comparison of varying complexity numerical models for the prediction of flood inundation in Greenwich, UK

T.J. Fewtrell & P.D. Bates
Willis Research Network, University of Bristol, Bristol, UK

A. de Wit & N. Asselman
Deltares/Delft Hydraulics, Delft, The Netherlands

P. Sayers
HR Wallingford Ltd., Wallingford, Oxfordshire, UK

The UK government estimates that in England and Wales alone, over 4 million people and properties valued at over £200 billion are at risk from fluvial, coastal and pluvial flooding. Furthermore, increased risk of flooding from climate change and associated sea level rise, increased rainfall and storm frequency mean that low-lying areas will be at greater risk from flooding in future years. Urban flood events contribute most to overall flood risk and thus there is a real and current need to develop computationally efficient urban flood models for risk analysis and wide area application. With the increasing availability of high resolution topographic and topological information for urban areas, in the form of LiDAR and MasterMap® data products, two dimensional (2D) flood inundation modelling of urban areas has become feasible.

This paper compares the ability of two numerical models of varying complexity to predict the evolution of a flood event in the Greenwich embayment of the Thames estuary. As part of the FLOODsite project, HR Wallingford derived probabilistic flood volumes for different event return periods and defence failure scenarios based on a probabilistic defence failure model. The derived flood volumes and defence failure durations were used to drive the flood inundation models, LISFLOOD-FP and SOBEK. LISFLOOD-FP is a 2D storage cell based numerical model developed at the University of Bristol whereas SOBEK is a fully hydrodynamic 2D model using an implicit ADI solver on a structured grid developed at WL Delft Hydraulics.

Recent modelling studies of urban floods have benchmarked models of varying complexity at a single model resolution finding that different classes of model produce plausible but subtly different model results. In order to establish a practical approach for developing urban flood models, it is necessary to assess the performance of models of different complexity at different resolutions. Therefore, model scenarios at grid resolutions of 5, 10, 25 and 50 m using a 1-in-100 year return period event with two defence failures were used to force LISFLOOD-FP and SOBEK. Furthermore, the appropriate choice of digital elevation data will be crucial in accurately resolving flood extents in dense urban networks for coarse resolution models.

Firstly, a comparison of the high resolution 5 m simulations in LISFLOOD-FP and SOBEK produces a coherent flood extent map with subtle differences relating to schematisation differences between modelling systems. The results suggest faster flood propagation in LISFLOOD-FP which leads to water surface height differences of the order of ~0.2 m in complex street networks. These differences may be caused by a combination of inertial effects not represented in LISFLOOD-FP and the schematisation of model inflows. Secondly, both models predict significant degradation in model results at coarse resolutions compared to the high resolution benchmark. Specifically, results at 25 and 50 m suggest different flow structures emerge as the representation of urban structures in the digital elevation models (DEM) becomes more coarse. As a result of this finding, the LISFLOOD-FP and SOBEK models are setup using a digital terrain model (DTM) with the buildings removed to investigate these emergentflowstructuresThepredictedfloodextentsfromtheDTMdrivenmodelssuggestthatanintelligentreprojectionmethodologymaybeanappropriatewayofaddressingemergentflowstructuresinfloodmodelsatcoarseresolutions.

Keywords: Digital elevation model, Urban flooding, Flood modelling, Flood risk management

Fast 2D floodplain modeling using computer game technology

R. Lamb, A. Crossley & S. Waller
JBA Consulting—Engineers & Scientists, Skipton, UK

1 NEEDS FOR 2D FLOOD INUNDATION MODELLING

In recent years, 2D flood inundation modelling has become an important part of flood risk management in practice, both for government flood management agencies and for the insurance and reinsurance sectors. Policy initiatives such as the European Floods Directive (and in the UK the national flood mapping programme) have created a strong demand for modelling flow over irregular floodplain topography to determine flood depths (and increasingly velocities).

2 COMPUTING CONSTRAINTS

There are several well established commercial codes now in routine use for 2D flood modelling, as well as numerous 'research codes'. Whilst the implementation details differ, most of the models used commercially solve the Shallow Water Equations, SWE (or approximations) by numerical integration. Despite advances in processing, it can still take a long time to set up and run a 2D model, which limits the number of scenarios, spatial extent or level of detail that can be delivered. In particular, the emergence of risk-based methods for flood modelling, which often involve Monte Carlo simulation of multiple event scenarios, demands intensive 2D modelling.

3 TECHNOLOGICAL ADVANCES

To meet the demand for faster 2D modelling, there have been some developments of quicker but very approximate codes, model emulators and also application of parallel processing techniques. For many businesses, the costs of 'true' supercomputing hardware and software are a significant barrier and so parallelism is typically exploited in a very simple way via grids of PCs or by multi-threading over the 2 or 4 core CPUs found in most workstation PCs. However, the multi-billion dollar market for games and computer graphics has driven the development of fast graphics cards (GPUs). Modern GPUs are, in effect, specialised high performance parallel processors with dedicated, high speed memory and currently up to 128 floating point arithmetic cores on one chip. These GPUs are available off-the-shelf with costs typically around €300 and will work in a standard workstation PC with very little customisation.

4 PERFORMANCE GAINS

We have written an implementation of a 2D flood model to exploit the parallel architecture of the GPU. As our starting point, we used the JFLOW diffusion wave model that has previously been applied to generate depth and extent grids for over 1,000,000 km of river in the UK. We present here benchmark results for the GPU implementation using data from the study of Hunter et al. (Proceedings of Institution of Civil Engineers—Water Management, In Press), based on an urban area of Glasgow, and compare them with results for other 2D models. Table 1 shows the run times for the original and GPU codes for a simulated flood event of two hours

Table 1. Performance of JFLOW versions using the urban benchmark test data from Hunter et al.

	Dual core CPU	GPU	Speed up factor
Run time (hours: mins)	41 hrs: 30 mins	0 hrs: 9 mins	~275

duration. We discuss the interpretation of the benchmark timings, comparison with 'full SWE' models and performance for other cases. Finally we report on the deployment of the GPU technology for broad-scale and high resolution flood modelling.

Keywords: 2D flood model, high performance computing, GPU, urban flood model, benchmarking

Grid resolution dependency in inundation modelling: A case study

S. Néelz & G. Pender
Heriot-Watt University, Edinburgh, UK

Ideally, detailed prediction of flood flows over complex topographies and in urban areas requires hydraulic model simulations on high resolution grids that can resolve the effects of individual buildings and other topographic features likely to have an influence on the movement of water across the flood plain. However, the associated computational cost may make the approach unfeasible where there is a need to simulate many flood event realisations (strategic risk planning), and/or where there is limited time or resources available to perform the computer simulations (real-time forecasting, emergency planning, studies with limited availability of staff or computer resources). In such circumstances, much coarser resolution models are often used to predict the general trends on the flood inundation. Such coarse grid models are informative, although to a lesser extent than fine grid models. There is therefore a trade-off between predictive capability and computational efficiency which needs to be addressed in a different manner depending on the intended application of the model.

The aim of this paper is to investigate how resolving the detail of the hydraulic processes depends upon model grid resolution using a hypothetical study case using a ~5 km^2 densely urbanised coastal floodplain. Inflows to the floodplain are assumed to occur from the failure of a flood defence during an extreme storm surge event. Results from a high-resolution 2 m square grid model are compared with 10 m and 50 m square grid model predictions for the same boundary conditions. Statistics are computed for a range of predicted variables, namely peak and final inundation extents, and inundation timing, final and peak water level and velocities at a large number (>1000) of output points across the floodplain. This is also done for a range of inflow magnitudes reflecting the uncertainty on this parameter. The dependency of detail prediction on grid resolution is investigated for each output variable, and is also compared to the dependency of prediction on inflow magnitude.

The main observations are that: 1) inundation extent is comparable between fine and coarse grid models; 2) for the timing of inundation and the peak flood level agreement is very good between the fine grid and the 10 m model, while with the 50 m model, although a degradation in the detail is observed this is less than that created by uncertainty in the inflow boundary condition; and 3) flow velocities are not well resolved with either the 10m or the 50 m model. A key conclusion is therefore that there are major differences in terms of sensitivity to resolution depending on whether flood levels or velocities are considered. Recommendations are made on the choice of model grid resolution for similar urban floodplains. The very high computational cost of predicting velocities is emphasised, considering that in practical applications the importance of predicting velocities accurately will need to be weighed against the necessity to investigate model uncertainty or/and to simulate a number of inflow scenarios.

Keywords: Inundation modelling, grid resolution dependency, uncertainty

2D overland flow modelling using fine scale DEM with manageable runtimes

Johan N. Hartnack, Hans G. Enggrob & Morten Rungø
DHI, Hørsholm, Denmark

The use of modern surveying methodologies such as LIDAR techniques, results in high resolution spatial data. The horizontal resolution is in many cases in the order of 1 meter or even less and with modern post processing techniques the vertical accuracy is in the order of a few centimetres. Such fine scale data sets seem ideal for 2-dimensional overland flow modelling; however modelling at such fine scale is often not practically manageable due to the long model run-times. Choosing the cell size of a 2D flood model thus often in practical cases becomes a compromise between accuracy and model run-times, which in turn means that topographical data has not being fully utilised.

To address this compromise a method has been developed which combines the benefits of a 2-dimensional overland solver with a technique for utilising of the finer scale topographical data. Simulation of the time varying water surface elevations and flow velocities is done on a grid which is coarser than the resolution of the topographical data. The fine scale topographical data is utilized when evaluating the governing equations for the 2-dimensional solver. In comparison with a standard 2-dimensional finite difference solution to the St. Venant equations, this method solves the governing equations with certain terms within the coarser grid solution modified with respect to storage area, flow area and hydraulic radius. Resulting flood depths and velocities are produced at the fine scale resolution. This paper describes the modifications to the solution to the continuity and momentum equations as well as MIKE FLOOD application examples showing pros and cons of using the new method. Amongst the pros are reductions in model runtime by a factor of 5–10.

Keywords: Flood modelling, fine scale topographical data, model run-times

Detailed 2D flow simulations as an onset for evaluating socio-economic impacts of floods

B.J. Dewals
Research Unit of Hydrology, Applied Hydrodynamics and Hydraulic Constructions (HACH) Department, ArGEnCo, University of Liege, Belgium & Postdoctoral Researcher of the Fund for Scientific Research, F.R.S.-FNRS, Belgium

S. Detrembleur, P. Archambeau, S. Erpicum & M. Pirotton
Research Unit of Hydrology, Applied Hydrodynamics and Hydraulic Constructions (HACH) Department, ArGEnCo, University of Liege, Belgium

In the framework of the Belgian national research project "*ADAPT—Towards an integrated decision tool for adaptation measures*", the hydrodynamic model WOLF 2D serves as a core component of a decision-support system (DSS). The tool is dedicated to the integrated evaluation of flood protection measures in the context of increased flooding hazard as a result of climate change. This DSS takes into consideration hydraulic, economic, social as well as environmental parameters.

Prior to assessing the effects of different flood protection measures, a preliminary step consists in determining how inundation hazard and its socio-economic impacts are likely to be affected by climate change. The present paper will describe how the hydrodynamic model is implemented within this integrated assessment of flood impacts, with a focus on the selection of relevant modelling scenarios and on the detailed simulation results for a specific case study located on the river Ourthe in the Meuse basin (Belgium).

The hydrodynamic model is based on the shallow-water equations solved by means of a finite volume technique. Multiblock Cartesian grids are exploited, enabling local mesh refinements, while an automatic mesh refinement tool embedded in the hydrodynamic model leads to significant reductions in computation time for simulating steady flows. Wetting and drying of computation cells is handled free of mass conservation error. A depth-averaged k-ε turbulence closure is also implemented and increases the potentialities of the model to represent real complex flows. The hydrodynamic model has been extensively validated by comparisons with observations during past major flood events.

Updated hydrological data, such as perturbation factors affecting peak discharges, are being processed in the framework of a parallel project (*CCI-HYDR*). Based on an analysis of the results of several climate models and on hydrological modelling, this work is inevitably affected by a significant level of uncertainty due to the models themselves and to the discrepancies in the scenarios used for running the climate models. Therefore, the subsequent hydrodynamic simulations are run based on four different scenarios (ranging between an optimistic 5% and an extreme 30% increase in peak discharge), in order to evaluate the sensitivity of the hydrodynamic results with respect to the uncertainty in input data.

The hydrodynamic simulations, conducted in the framework of the *ADAPT* project, enable to translate the modifications in the expected discharges into updated inundation extent and characteristics (water depth and flow velocities in the floodplains) for the considered reaches of River Ourthe. Since a high resolution Digital Elevation Models (laser altimetry) is available for the floodplains, computation grids as fine as 2 m by 2m are used for the hydrodynamic simulations. The simulations are conducted for two different return periods (25- and 100-year), likely to lead to complementary protection strategies.

It is found that the results can be very sensitive to the perturbation factor affecting the discharge, due for instance to the inefficiency of some flood protection structures (e.g. dikes) above a threshold value of discharge (design discharge). Moreover, the complexity of the flow fields represented on the inundation maps recalls the relevance of exploiting 2D flow modelling. The whole set of hydrodynamic results serves as an input for the subsequent socio-economic analysis of the impacts of flooding. The outcome of the integration of the present hydrodynamic results with the evaluation of economic impacts, as well as the social analysis, are presented in two companion papers submitted for the conference (Dewals, B.J., E. Giron, J. Ernst, W. Hecq, M. Pirotton: *Integration of accurate 2D inundation modelling, vector land use database and economic damage evaluation*; Coninx, I., K. Bachus: *Social flood impacts assessment. Towards a tool for decision makers*).

Keywords: Inundation modelling, digital elevation models, LIDAR data, finite volume, climate change

Ensemble Prediction of Inundation Risk and Uncertainty arising from Scour (EPIRUS): An overview

Qingping Zou & Dominic Reeve
Centre for Coastal Dynamics and Engineering, School of Engineering, University of Plymouth, UK

Ian Cluckie
Water Research Centre, Department of Civil Engineering, University of Bristol, UK

Shunqi Pan
Centre for Coastal Dynamics and Engineering, School of Engineering, University of Plymouth, UK

Miguel Angel Rico-Ramirez & Dawei Han
Water Research Centre, Department of Civil Engineering, University of Bristol, UK

Xin Lv, Adrián Pedrozo-Acuña & Yongping Chen
Centre for Coastal Dynamics and Engineering, School of Engineering, University of Plymouth, UK

1 INTRODUCTION

In the UK coastal flood defences are usually designed to withstand events with a return period of between 50 to 200 years, taking account of sea level rise. In 2006 the UK Natural Environmental Research Council (NERC) initiated a £5m research programme on flood risk under extreme events (FREE) to fund research into flooding arising from extreme events (> 50 year return period).

The interactions between atmospheric, oceanic and coastal processes are poorly understood, resulting in large uncertainties in the performance of sea defences and predictions of coastal flood risk in extreme conditions. NERC has funded the EPIRUS project to bring together a team of hydrometeorologists, oceanographers and coastal engineers to address this issue.

2 METHODOLOGY

This project consists of three models:

2.1 Meteorology model

The ensemble regional weather forecasting system will consist of the PSU/NCAR MM5/WRF and Met. Office UM mesoscale models and the global analyses/forecast datasets from the ECMWF. A dynamical-downscaling approach is being applied to resolve the dynamics over 1km grids. Based on these models and ECMWF ERA 40 Datasets, extreme weather futures will be generated with and without climate change and used to drive the coastal models.

2.2 Tide, surge and waves model

The regional POLCOMS offshore waves and tides model, the POL 2D tide-surge shelf-scale model and an existing 2-D high-resolution coastal process model are being used for transforming the surface meteorological fields into coastal parameters. The updated POLCOMS model includes coupled waves and currents. The coastal process models include wave refraction/diffraction, breaking, reflection, wind, tides, and wave-current interactions.

2.3 Surf zone model

We are using a state-of-the-art surf zone hydrodynamics model, based on Reynolds-averaged Navier-Stokes equations (RANS) to determine wave overtopping, breaking, turbulence and streaming arising from the wave and water level outputs from the tide/surge/wave models. This model includes a free surface tracking scheme using the Volume of Fluid technique. For this project, it will be extended to include predictions of beach morphology.

3 MODEL INTEGRATION

For each member of an ensemble of past/future storms events, the predicted wind and pressure fields by the meteorology model will be used to drive the wave/surge/tide models. These give forecasts of offshore wave and mean water level, which in turn are used to drive the surf zone model to predict the beach and structure response and to establish ensemble predictions of coastal flood risk arising from overtopping and scour.

Keywords: Flood Risk analysis, Extreme events, Uncertainty, Ensemble predictions

Flood risk assessment using broad scale two-dimensional hydraulic modelling – a case study from Penrith, Australia

H. Rehman
Golder Associates, Sydney, Australia

R. Thomson
Cardno, Sydney, Australia

R. Thilliyar
Penrith City Council, Penrith, Australia

Australian east and north coast is subjected to frequent flooding from rainfall-runoff and met-ocean processes. The landmark city Sydney, on the southeast coast of Australia, has a long history of flooding. Since European settlement, a large number of significant flood events have been recorded in the three major river systems within the Sydney Metropolitan Area. Penrith is a large local government area (LGA) in the west of Sydney Metropolitan Area and lies within the catchment of Hawkesbury-Nepean river system.

Under the New South Wales State government guidelines, Penrith City Council has initiated the floodplain risk management process in the Penrith LGA. This process generally requires detailed flood studies to be undertaken followed by a comprehensive floodplain management options study, where various flood modification, property modification and response modification options are assessed using economic, social and environmental indicators. The final stage of the process is to prepare a Floodplain Management Plan that facilitates implementation of various floodplain management options.

Generally the floodplain management process is carried out on a localised scale for a small sub-catchment of the larger system. Penrith LGA has potentially 30 sub-catchments, which are a candidate for a detailed study.

Due to funding constraints, the detailed studies can only be carried out for a few sub-catchments at a time and therefore Penrith City Council needs to prioritise the sub-catchments on the basis of flood risk associated with each sub-catchment. To achieve this objective, Penrith City Council commissioned a broad-scale flood risk assessment study for the entire LGA. Given the scale of the study area (407 km^2), a novel modelling approach was adopted whereby topographic grids of suitable sizes were generated for the entire LGA. These grids were then used to undertake two-dimensional hydraulic modelling using 'rainfall on grid' approach, which utilises rainfall as direct input to the hydraulic model and eliminates the need to undertake the complementary hydrological modelling required for such studies.

Based on the topographic grids, a number of coupled one and two-dimensional (1D/2D) hydraulic models were set-up to represent two dominant land-uses within the LGA. For urban areas, 3m grid models were developed whereas for rural areas 9m grid models were established. Modelling was undertaken for a number of design flood events and modelling results were processed to generate flood extents, flood hazard extents along with discharge data at various locations in the study area.

Finally, modelling results were used to establish a risk factor for each sub-catchment to help prioritise the sub-catchments for future detailed studies.

This paper provides a brief overview of the current floodplain management process in New South Wales. Details of the modelling methodologies including 'rainfall on grid' approach are also provided. The paper also describes the process of flood risk analysis used to prioritise sub-catchments in the study area.

Keywords: Flood Risk Assessment, Modelling

Modelling and analysis of river flood impacts on sewage networks in urban areas

A. Kron, P. Oberle & A. Wetzel
Institut für Wasser und Gewaesserentwicklung, Universitaet Karlsruhe (TH), Germany

N. Ettrich
Fraunhofer Institut für Techno- und Wirtschaftsmathematik, Kaiserslautern, Germany

1 MOTIVATION

Hydrodynamic modellings in urban areas infer tougher challenges than those in rural areas. They require considering mega structures, highly varying land uses, intricate flow paths, and underground drainage networks. These tasks are not only time-consuming but also demanding lots of information and data. Because of geometric complexity of structures and storages in the building etc, to model the overland flows in urban area is by far still immature, not to mention the comprehensive coupling between sewerage networks, overland flows or river overflows. Floods in urban areas can be triggered by the following mechanisms: overflows from rivers, invasions of the land drainage flows from upstream hills, intense storms in local areas, and drainage networks surcharge or failure. These mechanisms can happen respectively, but most of the time they occur and interact simultaneously with each other. This research will provide a relatively comprehensive vision such that more than one of the above mechanisms are treated by the model.

2 STUDY AREA

Sited in southwestern Germany and the northern corner of the Neckar basin, Heilbronn is an intermediate city with approximately 120,000 residents and an area of 100km2. Although there exists an integrated flood protection measures in the city Heilbronn, she suffers the flooding from the river Neckar or the surcharge of sewage networks when extreme storm events occur. Our studied area is located in the northern part of the city. The docks, freight stations, power plant, wastewater treatment plant, depots, and residential districts are distributed therein. Furthermore, the river Neckar passes through the western and northern banks of this region. Owing to the characteristics of the studied region, such as isolated terrain, moderate spatial scale (2km2), highly varying land uses as well as influences of rivers and sewers, it is apparently an ideal region to perform simulation researches.

3 METHODS

Modelling of the overland flows that are caused by floodwater of underground drainage networks has become increasingly vital for the assessments of urban flood risk. Model coupling between different orders of dimension, say coupling of 1-D sewage networks and 2-D overland flows, is one of the state-of-the-art approaches. Not only the high-resolution terrain data and detailed information do we require, but also a sophisticated simulation model is even more essential for the sake of achieving the computation. This research employs the information and results of the project IKoNE (Integrierenden Konzeption Neckar-Einzugsgebiet), proceeded by the IWG (Institut für Wasser und Gewässerentwicklung) an der Universität Karlsruhe. Furthermore, it also makes use of the mature hydrodynamic simulation model, RisoSim, developed by the ITWM (Institut Techno und Wirtschaftsmathematik) Fraunhofer, to perform the researches including 1-D sewer networks modelling and 2-D surface floods as well as the both coupling. Our research concentrates on the impacts of overtopping from river flows; meanwhile, the interactions between sewerage networks and flooding caused by river flows are also performed. In addition, the up-to-date as well as high-resolution orthophotos (10cm) and DTM (Digital Terrain Model) are also used therewith. By means of comparing the results with simulations and measurements, the model is calibrated and validated; meanwhile, the model sensitivity analysis is performed as well. In the end, the urban potential flooding areas and the risk assessment are implemented by simulating different scenarios.

Keywords: Flood modelling, Flood analysis

Coastal flood risk modelling in a data rich world

R.D. Williams & M.R. Lawless
Jeremy Benn Associates Ltd., Skipton, North Yorkshire, UK

J. Walker
Environment Agency, Preston, Lancashire, UK

Technological and methodological developments have facilitated the acquisition of unprecedented amounts of high-resolution data on many aspects of the coastal environment. Sonar and airborne LiDAR surveys enable the development of detailed 3-D Digital Terrain Models (DTM); high temporal resolution tide and wave gauges provide information on sea state; and satellites monitor meteorological conditions and provide information on land use. This paper explores how the use of these datasets in the Environment Agency's North-West Central Area, together with sophisticated 2-D modelling, has generated new information on (i) extreme sea levels, (ii) extreme flood inundation extent, and (iii) the impact of environmental change. The availability of this information considerably improves the evidence available for policy decision making.

1 EXTREME SEA LEVEL MODELLING

Historic sea level information is the primary data required to estimate extreme sea levels accurately. Traditionally, this data has been obtained from tide gauges. However, gauge data are sparse in the NW. To overcome this obstacle, a long-term hindcast surge model was developed for the region using detailed meteorological and bathymetric data. The output from the model is a 'virtual gauge network', including a 43-year record of tide and surge levels for more than 5,000 sites. Extreme sea levels and design events were derived from this data using advanced joint probability techniques. This has provided a unique picture of how extreme sea levels vary in complex tidal regions. The approaches used will inform the new Defra/EA Science project (SC060064) on Coastal and Estuary Extremes.

2 TWO-DIMENSIONAL FLOODWATER INUNDATION MODELLING

Design events obtained from the Extreme Sea Level modelling were used to define the water level boundary condition for floodwater inundation models. The models were run using TUFLOW, a fully 2D model. The model topography was built from LiDAR data and defence levels were estimated from 0.25m horizontal resolution LiDAR. Calibration of the models was possible using gauge data along major estuaries. Wave heights were estimated from a SWAN model and overtopping discharges were calculated using the new techniques available from the European Wave Overtopping Manual. This approach has produced new flood inundation maps with unprecedented accuracy and detail. The 2-D models have also been modified to evaluate the opportunities and risks associated with defence realignment at Hesketh Out Marsh, on the Ribble Estuary, and to evaluate the impact of environmental change (for example, by investigating sea level rise processes and the impact of increased coastal erosion on wave dynamics at Sunderland Point).

3 BETTER DECISION MAKING

In NW Central Area, which covers Lancashire and part of Merseyside, approximately 70,000 properties are at risk from flooding under an extreme tidal flood event. Therefore, accurate modelling of the extent of the coastal floodplain is essential to good planning and flood risk decision-making. The methodology described above has the potential to significantly improve the accuracy of Environment Agency Flood Map outlines in this Area. The primary purpose of the Flood Map is to guide development by offering information on the location of natural floodplain areas, hence it shows the risk of flooding without defences. However, Local Authorities, the Environment Agency and others have previously invested heavily in coastal flood defences. Therefore the modelling approach described will also map the areas benefiting from these defences, and this will be added to the Environment Agency Flood Map for the first time. Additionally, this approach will be used to map increased risk from coastal flooding attributable to climate change.

Keywords: Extreme estimation, risk management, tide-surge, two-dimensional modelling

A multi-scale modelling procedure to quantify effects of upland land management on flood risk

H.S. Wheater, B.M. Jackson, O. Francis, N. McIntyre, M. Marshall & I. Solloway
Imperial College London, UK

Z. Frogbrook & B. Reynolds
Centre for Ecology and Hydrology, Bangor, USA

Recent UK floods have renewed speculation about the linkage between agricultural land management and flooding. Available data to quantify effects of agricultural intensification have been limited, small scale, and mainly focused on the lowlands and arable agriculture. There is a need to quantify impacts for upland areas, which are source areas for runoff generation, and to develop methods to extrapolate from small scale observations to predict catchment-scale response. With assistance from a cooperative of Welsh farmers, and support from the EPSRC Flood Risk Management Research Consortium, a multi-scale experimental programme has been established at Pontbren, in mid-Wales, an area of intensive sheep production. The data have been used to support development of a multi-scale modelling methodology to assess impacts of agricultural intensification and the potential for mitigation of flood risk through land use management.

Data are available from statistically-replicated experimental plots under different land management treatments, from instrumented field and hillslope sites, including tree shelter belts, and from first and second order catchments. Measurements include rainfall and climate variables, soil moisture, soil water pressure and soil hydraulic properties at multiple depths and locations, tree interception, overland flow and drainflow, groundwater levels, and streamflow from multiple locations. Detailed, fine resolution, physics-based models have been developed to represent soil and runoff processes, and conditioned using experimental data. The response of these detailed models is used to develop and calibrate simpler 'meta-models' to represent individual hydrological elements—in this case mainly individual fields, with their associated field drainage. These meta-model elements are then combined in a distributed catchment-scale model.

The paper presents results of detailed field-scale simulations to demonstrate the dominant runoff processes under intensive sheep production, and impacts of the use of tree shelter belts in improving soil structure and reducing peak runoff intensities. Catchment-scale simulations show the effects of improved and unimproved grassland, and the potential effects of land management interventions, including farm ponds, and tree shelter belts and buffer strips. It is concluded that the methodology developed has the potential to represent and quantify catchment-scale effects of upland management; continuing research is extending the work to a wider range of upland environments and land use types, with the aim of providing generic simulation tools that can be used to provide strategic policy guidance.

Key words: Land use, land management, runoff processes, flooding

Updating flood maps using 2D models in Italy: A case study

F. Nardi
Honors Center of Italian Universities H2CU, Rome, Italy
Hydraulics Applied Research and Engineering Consulting s.r.l., Harec, Rome, Italy

J.S. O'Brien
FLO-2D Software inc., Nutrioso, Arizona, USA

G. Cuomo
Engineering and Hydraulic Structures Group, HR Wallingford Ltd., Wallingford, UK
Hydraulics Applied Research and Engineering Consulting s.r.l., Harec, Rome, Italy

R. Garcia
FLO-2D Software inc., Nutrioso, Arizona, USA

S. Grimaldi
Gemini Department, Tuscia University, Viterbo, Italy

River basin authorities in Italy have recently published new political and technical standards that require updating flood zoning maps by applying two-dimensional flood models. Two-dimensional simulations represent an effective tool for flooplain managers to improve the prediction and management of flood hazards for those areas where the loss of channel conveyance results in large unconfined overland flow with extensive floodwave attenuation.

A review of existing regulations in Italy is presented that focuses on the comparison of the different river basin authority standards in relation to the implementation of 1D versus 2D numerical models. A real case study is presented, involving Lazio Region Basin Autorithy, in which the original flood maps, based on the implementation of a 1D river flow simulation, have been updated by using results of a 2D flood model ultilizing the new flood map regulations.

1 THE NUMERICAL MODEL

Numerical simulations have been carried out using the two-dimensional FLO-2D model (www.flo-2d.com). FLO-2D is a volume conservation model that numerically routes a flood hydrograph while predicting the area of inundation and floodwave attenuation over a system of square grid elements. The model considers river channel flow as one-dimensional and simulates river overbank flow and floodplain flow as two-dimensional. In this case, the model was used to simulate unconfined flow on complex floodplain topography and roughness in the vicinity of the Rio and Eri Creek confluence, The FLO-2D features used on this model also included levees, hydraulic structures, and the effects of flow obstructions. Flood routing in two dimensions was accomplished through numerical integration of the equation of motion (full dynamic wave momentum equation) and the conservation of fluid volume.

2 CASE STUDY: UPDATING FLOOD ZONING MAPS OF THE RIO AND ERI CREEKS

The FLO-2D model have been applied to the real case of the Rio creek and its tributary Eri flowing into the Tyrrhenian Sea, just North of Rome. HEC-RAS one-dimensional single-discharge model computations are compared to FLO-2D results routing the 200 years return period flood hydrograph. This project demonstrates how the channel-floodplain flow interchange, the floodwave attenuation, the unconfined flow on the floodplain generated by levee overtopping, simulated by FLO-2D result, in this case study, in a significant improvement in

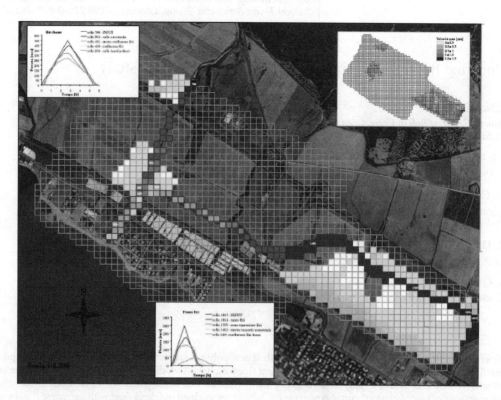

Figure 1. Numerical simulation of the 200 years return period flood using FLO-2D for the Rio and Eri Creek in Central Italy.

the understanding of the flood inundation process as respect to HEC-RAS that consistently overestimated the inundation area. FLO-2D project results have been submitted and approved by the Technical Committee of the Lazio Region River Basin Authority for updating the Rio-Eri river system flood maps.

Keywords: Floods, Two-dimensional 2D flood model, Flood mapping, River Basin Authority, Italy

Real-time validation of a digital flood-inundation model: A case-study from Lakes Entrance, Victoria, Australia

P.J. Wheeler, J. Kunapo, M.L.F. Coller, J.A. Peterson & M. McMahon
Monash University, Victoria, Australia

Extensive flooding in the Gippsland Lakes catchment (Victoria, Australia) during June/July 2007 caused much damaging urban area inundation within townships located on subdued, low-lying areas along semi-enclosed coastal lagoon shores. Data capture during the 2007 flood event facilitated 'real-time' validation of a photogrammetrically-derived high-resolution bare-earth terrain model, suitable for flood and storm surge/storm tide inundation modelling, which was independently developed for the townscape of Lakes Entrance, Victoria, Australia, by Wheeler et al. (2007). A lack of corporate and third-tier stakeholder management flood-risk digital spatial decision-support characterised this flood event, despite these stakeholders experiencing a similar serious flooding event taking place some nine years previously (in June/July 1998). The results presented here indicate the immediate need for deployment of 'detailed geography'; namely, the high-resolution digital elevation modelling of low-lying townscapes located along the shores of the Gippsland Lakes. It is argued that this type of modelling, and subsequent spatial data dissemination and stakeholder access (via regional SDI or issues-based web-map server) is essential to provide disparate public and private management agencies with an integrated spatial decision-support capacity *vis a vis* future flood inundation risk, planning and management, in support of integrated coastal zone management (ICZM) initiatives.

Keywords: Digital elevation model, digital photogrammetry, coastal lagoon, coastal inundation modelling, ICZM

Dispelling the myths of urban flood inundation modelling

David Fortune
Wallingford Software, HR Wallingford Group, Wallingford, Oxfordshire, UK

For operational purposes, most flood inundation modelling is carried out using 1D and pseudo-2D methods. Yet we are all aware of the relative benefits of 2D flow modelling. Of course, academia has been experimenting with 2D solutions to the shallow water equations for some years. But very little of this work has found its way into "real" life.

When it comes to simulating urban flood flows, people are generally more cautious, with good reason. Most widely available 2D models can be made to work reasonably at low levels of detail for urban areas, but the level of effort involved in building and tuning these models is usually prohibitive for all but the most urgent cases. And for complex flow paths, even well accepted 2D models have struggled to match known flood routes. Many distressing, repetitive, flood problems have gone unsolved because water managers have not had the tools available to quickly and simply analyse the source and pathways of localized flooding.

We all know that urban flooding is a complex issue, with potential sources such as rainfall runoff, drainage overload and failure, rural runoff, groundwater, urban rivers, wave overtopping and surge. Modelling can be a great tool for analysing any of these flood sources individually, but water managers generally give-up on modelling when one or more sources is to blame simultaneously. There has been much talk of integration, but in practice, we have generally avoided the issues of simulating flow between these various domains, prefer to spend considerable time constructing assumed boundary conditions between them.

So for post-flood analysis, we have rightly put very little faith in modelling solutions. Even where we have got the right tools in theory, using them in practice has generally been too complex, taken too long and been too costly. For real-time flood forecasting, the received wisdom is clear: stay away from 2D flow modelling, the run times are too long. And nobody would dream of using 2D models for rainfall runoff in real-time.

Well many of the "truths" stated above are no longer the case. They are myths. The natural way to simulate urban flood flows will soon be with 2D models, even for real-time flood forecasting. What has changed? For a start, readily available computer power has increased to a suitable level. But this is only part of the answer. It is also true that 2D algorithm implementations have improved in recent years, following on from much valuable research for the avionics industry. Of course, in some parts of the world we suddenly have much more good terrain data available at acceptable prices from LIDAR and other sources. The practice of integrated modelling has been advanced by the advent of the OpenMI standard for data exchange. But even when not using OpenMI, software developers have found acceptable ways to integrate models on a piecemeal basis. And industrial modelling systems have improved in efficiency and quality control beyond all expectation. Modelling has gone from a costly, niche, academic activity to become a "standard" part of water management.

This paper illustrates the way that research and industry has worked in partnership to bring these advances together, to create a new paradigm in urban flood inundation modelling. We can dispel the myths and look forward to widespread use of flood modelling in water management, even for real-time flood forecasting!

Keywords: Inundation, urban flood modelling

Flood risk in urban areas caused by levee breaching

André Paquier & Charly Peyre
Cemagref, Lyon Cedex 09, France

Nicolas Taillefer & Menad Chenaf
CSTB, Marne la Vallée Cedex 2, France

Urban areas are often protected against floods by levees set along the riverbanks. Then, a flood becomes a rare event due to either a very high flood or a failure of the levee system. In the latter case, the flooding process can be very sudden thus preventing any easy evacuation of the inhabitants. Then, some question stand: will people be safe in the buildings? Will buildings resist to the flow? How fast the water will enter the buildings?

According to the type of building met (construction design, number of levels, etc) and the type of event that can damaged the building (static or dynamic pressure on the walls, doors and windows, entrainment of debris such as trees and cars, erosion), an assessment of vulnerability is proposed that links the possible maximum damage to a couple of hydraulic variables (water depth and velocity). It is discussed if peak water depth and peak velocity are enough to conclude.

In France, for PPRI (Plans for Prevention for Risk of Inundation), one single flood has to be considered: either the 100-year flood or the highest historical flood if higher. Then, hydraulic variables are calculated for this single event and for a levee breaching that is generally difficult to locate. In this latter case, numerical modelling should consider any location of the breach in order to estimate the situation with higher risk. Consequently, the more important parameter is often the shorter distance to the levee system.

Because the failure of one levee is a very rapid process (generally a few minutes), numerical modelling should include the whole system: flow in the river, levee and protected flood plain. A lot of calculations (linked to breach locations) should be performed. Thus, engineers would prefer using a simplified propagation model to replace the classical 2D model (solving shallow water equations). Results applying simplified CastorDigue software are compared to results applying the classical 2D software Rubar20. Both codes include a specific model to simulate the erosion of an embankment; thus, the discussion is focussed on the flow propagation in the flood plain. Special attention is paid to the influence of buildings (location, density, ...). Field cases (Agly River, Lez River, ...) are used to illustrate various situations particularly concerning the duration of the flood.

Finally, it appears that, if mitigation measures are concerned, detailed 2D calculations are necessary while, for emergency measures, only simplified calculations are enough.

Keywords: Levee breaching, urban flood, numerical modelling, planning

RISK-EOS flood risk analysis service for Europe

V. Holzhauer & M. Müller
Infoterra GmbH, Friedrichshafen, Germany

A. Assmann
Geomer GmbH, Heidelberg, Germany

Since it started in 2003, RISK-EOS is a network of European service providers delivering geo-information services to support the management of flood, fire and other risks throughout all phases of the risk cycle (prevention, early warning, crisis and post crisis). The RISK-EOS services combine the use of satellite observation data with external data and modelling techniques. They serve the needs of all risk management stakeholders at European, national and regional levels, in particular civil protection, fire fighting and rescue services, spatial planning, risks prevention services and territorial communities.

The RISK-EOS services have been developed and qualified in close cooperation with operational users from different European countries. RISK-EOS is part of the GMES Service Element Programme and is supported by the European Space Agency. GMES—a joint initiative of the European Commission (EC) and the European Space Agency (ESA)—aims at designing and implementing a European capacity for the provision and use of operational information for Global Monitoring of Environment and Security (GMES).

This paper will concentrate on the flood risk analysis portfolio:

Flooding is the most frequent natural disaster affecting millions of people in Europe and representing over 75% of natural disasters occurring worldwide. It has an enormous impact on the environment and society, causing pollution by release of hazardous substances into surface and groundwater and massive destruction of important infrastructure. Coping with floods has been mainly seen from a technical point of view towards flood protection. In recent years the focus has shifted towards an integrated flood risk management including potential risk modelling as well as cost and benefit calculations.

The flood risk analysis service provides the creation and maintenance of geo-information for areas which are potentially subject to flooding. The main objective is to support decision making in flood risk management. The products include information on past and potential flood events as well as calculations of potential damages and economic losses. An important aspect and benefit of the RISK-EOS services is the combination of satellite based information with high performance modelling techniques and detailed ground data. The results are risk maps displaying the current situation. The integration of recent data into a flood information system (www.floodrisk.eu and local user systems) allows for an efficient and supportive use of the results. The products, such as past flood event maps, flood risk maps and damage assessment information are distributed upon user requests. The services and products are compatible to the requirements of the EU flood directive and already meet most of its requirements.

Based on many years of user feedback and testing of the maps in real events, the production process and scenario definitions have been optimized towards the user needs. It will be shown that the scenario definition and the modelling technique have a very significant impact on the resulting risk maps. The experiences made will help to find the best fitting scenario definitions for each type of river. Also open issues on this topic (like volume of hydrograph) are discussed.

Furthermore, additional requests like dynamic maps (including flood duration, extent at different time steps, flow velocities and direction) and simulation of failure scenarios of flood retaining structures are fulfilled by the services. The cartographic representation also plays an important role in making the information useful to the end user(s).

Keywords: GMES, RISK-EOS, modelling, flood risk maps, EU flood directive, flood information, FloodServer

Flood inundation modelling: Model choice and application

N. Asselman, J. ter Maat, A. de Wit & G. Verhoeven
Deltares/Delft Hydraulics, Delft, The Netherlands

S. Soares Frazão, M. Velickovic, L. Goutiere & Y. Zech
Fonds de la Recherche Scientifique—FNRS and Department of Civil and Environmental Engineering, Université catholique de Louvain, Louvain-la-Neuve, Belgium

T. Fewtrell & P. Bates
School of Geographical Sciences, University of Bristol, Bristol, UK

Flood inundation models are essential in flood risk management. They provide information on expected inundation patterns, including water depths, flow velocities and timing of inundation. This information is needed for the preparation of evacuation plans, and for the development of sustainable flood risk management polices and intervention measures.

Different types of inundation models exist, varying from very simple models, based on the extrapolation of observed water levels, to more complex 2- or 3-dimensional models. For non experienced modellers it often is difficult to determine what type of model they should apply and how to apply it. Questions that arise concern the grid cell size to be used in case of a 2D-model and the processes that should be included (such as breach growth, wind effects and evaporation).

Within the **FLOODsite** project, a comparison of different modelling techniques and options was carried out with the aim of defining best practice guidelines. A series of simulations was carried out for the Thames and the Scheldt estuary, investigating the consequences of, among other things, the type of spatial discretisation (square grid vs. triangular grid), grid resolution, the influence of the breach formation process and schematisation, the influence of wind, and the schematisation of built-up areas.

Three different numerical models were compared: (1) SOBEK developed by WL/delft hydraulics (now Deltares/Delft Hydraulics), (2) SF2D, developed at the Université catholique de Louvain and (3) Lisflood-FP developed by the University of Bristol.

The results indicate that SOBEK 2D, LISFLOOD and the UCL model are all 2D models that are suitable for the simulation of flooding in areas similar to the Scheldt and the Thames pilot sites. However, at locations where inertia plays an important role, LISFLOOD might be slightly less suitable than the other models, as LISFLOOD does not account for this. Application of 1D models in quasi-2D mode only produces accurate results when the flow pattern is known in advance. If the 2D flow pattern is not known, it becomes very difficult to represent areas with complex topographies (such as urban areas) in an accurate manner. This results in large errors.

The results also indicate that when a flooding model is developed for a mainly urban area (i.e. a large part, if not the entire model area is urbanised), the maximum grid cell size should not exceed the gap distance between buildings (typically 10 m for many European cities). In more rural areas where villages occur locally, a much larger grid cell size can be applied as long as one is interested in the flooding characteristics over a larger area. Flow velocities are not computed accurately if built-up areas are represented by increased roughness values instead of solid 2D objects and a finer grid cell size. Storms can have a large impact on computed water depths and inundation patterns. In case of the Scheldt model, differences of more than 1 m are observed. Variations in hydraulic roughness also have an effect on the model results. The impact of other model choices (grid cell size, schematisation of buildings, etc.) and processes (e.g. wind) is, however, much stronger.

Keywords: Flood models, modelling techniques, bench mark, resolution

Risk maps of torrential rainstorms

A. Assmann
Geomer GmbH, Heidelberg, Germany

M. Krischke & E. Höppner
Intermap Technologies GmbH, Munich, Germany

In addition to large river floods it is more and more recognized that the large number of local torrential rainstorms cause economic damages in the same order of magnitude. This can be learned from looking into historical archives. However, the possibilities of forecasting these events are still very limited and the assessment of the temporal and spatial distribution of potential risks associated with torrential rain events is hardly ever performed.

These risks are governed by many factors like the amount of precipitation, soil type, underlying geology, typo of relief and land use patterns. The smaller the catchment the more the runoff generation process will determine the discharge. With increasing catchment size retention and hydraulic effects of the drainage system become more important.

As this areas lay outside the responsibility of the EU flood directive still also in the future about 50% of the flood damages will occur without having the people informed or warned in advance.

Additional to the runoff generation process the relief has a very high impact on the concentration and back-flow processes. For determining the areas at risk it is very important to have high quality terrain data with a resolution of at least 5m and to use a two-dimensional hydraulic model.

As equivalent data are becoming available by now it is possible the first time to produce risk maps for torrential rain. Intermap's high resolution radar DTM provides a data set with a homogeneous quality for most of Europe in the near future. This data set allows for the production of homogenous and thus comparable results all over Europe.

Now it is the main challenge to handle this large amount of data while modelling at a resolution of 5 m or below. For example: just for the area of Germany a number of 15 billion calculation nodes need to be processed.

By introducing computer clusters and new software (like FloodArea HPC—High Performance Computing) to the simulation process it is possible to handle this enormous amount of data within a detailed precipitation and runoff model.

The resulting maps display the areas affected by the runoff of torrential rainstorms. Also information about the maximum flow velocities can be generated, an important information concerning soil erosion associated with such rainfall events. The maps are a perfect basis for planning authorities, civil protection and rescue services, architects, and home owners. In contrast to the large river floods, the prevention measures need to be organized at a very local level. Luckily the costs are limited in most cases. However, the planning of prevention measures is only possible if proper information about the sources, the flow paths, and the amounts of the torrential waters are available.

An additional benefit of this modelling process is the derived information about areas where most parts of the flood's runoff are actually generated. This information serves as the basis for planning decentralized flood protection measures.

The first risk maps have been produced within the INTERREG IIIB-project AMEWAM, currently simulation test are performed on the scale of a German federal state (850 million raster cells in a single simulation run). Depending on the demand, one or two European countries should be processed until the end of 2008. Such risk maps are requested by insurance companies, planning agencies, communities but are also very important for private house owners to get prepared for the next torrential rainstorm.

Keywords: torrential rainstorm, modelling, flood risk maps, AMEWAM, decentralized flood protection

Decision Support System for flood forecasting and risk mitigation in the context of Romanian water sector

I. Popescu, A. Jonoski & A. Lobbrecht
UNESCO-IHE Institute for Water Education, Delft, The Netherlands

The Netherlands has a long history of water management, during which knowledge and experience have been gained on how to protect the country, its inhabitants and other water uses such as industry, agriculture and nature, against floods. In particular the knowledge and experience in the field of forecasting of excessive rainfall events, forecasting and control of high river water levels, as well as mitigation of floods, can be exported to other countries both in Europe and to the rest of the world. Like the Netherlands, Romania is a country where rivers discharge into the sea. This brings particular problems to water management such as end-of pipe effects in terms of water quantity and water quality. Romanian Waters (The National Agency responsible for overall water resources management) follows the legislation compatible with the EU regulations regarding water resources management and the preservation of aquatic ecosystems and water areas. In this respect, this agency is responsible for the ways in which the surface and underground waters on the Romanian territory are used.

In terms of flooding problems one of the most vulnerable regions in Romania is in the West. Furthermore, many rivers in this region are of transboundary nature. Most of the rivers from the western side of Romania are having their basins in either Romania and Serbia, or, Romania and Hungary. Any event occurring on these rivers is advected downstream to the neighbouring country. A typical example of this situation is the river Timis, which in recent past caused some severe flooding in the two neighbouring countries of Romania and Serbia. In the past many dikes have been constructed along the river for flood protection, which in return made the downstream flood propagation quicker, causing severe flood damages.

Within a framework of a collaborative project between Romania and the Netherlands a demonstrator of a flood forecasting system was developed, which can support operational water management under extreme conditions when actions have to be taken quickly. The pilot basin for the development of the system was the Timis-Bega river basin, in which the rivers Timis and Bega were considered jointly, since their hydrodynamic responses are conditioned by operation of existing hydraulic structures used for water transfer between the two rivers. The main use of the system is in forecasting rising and high river water levels, while in future it can be extended to prediction of dike bursts which will enable taking of proper and timely preventive and mitigation measures.

The system is based on a combined hydrological and hydrodynamic model, developed specifically for the pilot region. Hydro-meteorological data such as rainfall, temperature and water levels are automatically retreived from existing stations, and they are fed into a hydrological model that is run in forecasting mode. The retrieved data, together with the provided forecasts are deployed on a dedicated web-site, which can be used by the operational managers. The system is developed with the modelling systems HEC-HMS (hydrological model) and HEC-RAS (hydrodynamic model). The work is currently extended into coupling the HEC-RAS model with a 2-D SOBEK model for downstream flood analysis and risk assessment, including dike breach scenarios.

Keywords: Flood risk management, Decision Support Systems

Developing a rapid mapping and monitoring service for flood management using remote sensing techniques

Vasile Craciunescu, Cristian Flueraru, Gheorghe Stancalie & Anisoara Irimescu
National Meteorological Administration, Bucharest, Romania

Flooding remains the most widely distributed natural hazard in Europe, leading to significant economic and social impact. The remote sensing of the Earth is presently capable of making fundamental contributions towards reducing the detrimental effects of extreme floods. Both optical and radar wavelengths will increasingly be used for disaster assessment and hazard reduction. These tasks are accomplished, in part, by accurate mapping of flooded lands, which is commonly done over periods that can vary from days to months.

In the latest years, river flooding occurred quite frequently in Romania, affecting wide areas of the country's territory. That was the case in 2005 when severe flooding affected the Banat area (south-western part of Romania) in April and Moldova area (eastern part of Romania) in July. The effects: more than 20 people killed, hundred of houses destroyed and thousand of hectares of land flooded. The scenario repeated in spring 2006 when the southern part of the country was hardly hit when the Danube River burst its banks due to the heavy rain and huge amount of melting snow on Central Europe.

In this context the bases for a flood mapping and monitoring service were established. For all reported events, the remote sensing data played a major role in producing high temporal resolution flood maps. One of the key elements in producing such maps is the time spanning. A useful product should reach the decision makers in less than a day or two. That was possible by using images from Moderate Resolution Imaging Spectroradiometer (MODIS) on NASA's Terra & Aqua satellites. A dedicated methodology was developed and tested, in order to process the raw .hdf MODIS data, extract and integrate the water mask into a useful, standardized, cartographic product. Beside the static flood maps, 3D flythrough and different animations were produced to get a more complex perspective on flood dynamic and dimension. Damage assessment was performed by integrating the flood mask, in a GIS environment, along with relevant info-layers (land cover/use, transportation, populated places).

Parallel with the daily flood map production, GPS field measurements were carried out, to ensure the precision of the satellite flood mapping. Supplementary calibrations were performed using medium and high resolution satellite images (ASTER, SPOT).

The results were quickly disseminated, in digital form via Internet, using a dedicated webpage (http://www.inmh.ro/images/Floods/indexcfe0.html), and in printed form for the field intervention teams. The main core of users for these products was formed by the Romanian Ministry of Environment and the General Inspectorate for Emergency Situations, along with different institutions at local level.

Keywords: Rapid mapping, remote sensing, flood management

# A framework for Decision Support Systems for flood event management – application to the Thames and the Schelde Estuaries

D.M. Lumbroso
HR Wallingford, Wallingford, Oxfordshire, UK

M.J.P. Mens & M.P. van der Vat
Deltares/Delft Hydraulics, Delft, The Netherlands

There is a clear need for flood event managers to be able to improve the coordination of their response to a flood event. The development of a flood event management Decision Support System (DSS) could assist flood event managers in this by providing information on the flood hazard, the receptors at risk and evacuation routes. This paper describes the development of a general framework for DSSs for flood event management. This framework has been piloted in the Thames Estuary in the UK and the Schelde Estuary in the Netherlands, resulting in the development of two prototype DSSs.

Consultation with end users in the UK and the Netherlands indicated that a DSS for flood event management should provide the following: A database of "pre-run" flood events; Mapping of the flood hazard; Location of the receptors at risk (e.g. people, buildings, critical infrastructure); Details of safe havens and evacuation routes. Based on the user requirements a methodological framework for flood event management DSSs was developed. This framework includes exposure, vulnerability, consequence and risk modules. This framework was used as a basis for prototype DSSs developed for the Thames and the Schelde Estuaries. The two DSSs allow users to compare hazards and risks related to flood event management, using the outputs from hydrodynamic models.

For the Thames Estuary a Flood INcident Tactical and Operational Framework (FLINTOF) was developed and applied to the Thamesmead embayment. FLINTOF includes the following functionality: Estimation of the flood risk to people in terms of number of injuries and fatalities; Assessment of the road network with respect to emergency access; The display of typical evacuation times at a census enumeration level; Estimates of the probability of building collapse; The appraisal of different emergency management interventions.

The FLINTOF system allows flood risk managers to go some way to addressing key flood event management issues. The Environment Agency has indicated that they need a systems based approach to emergency management. Consultation with end users indicated that spatio-temporal risk estimation has strong potential to improve emergency response. The FLINTOF system is flexible and could be applied throughout England and Wales. In future the rapid risk estimates provided by FLINTOF could be integrated with other real-time information technologies to produce an integrated real-time emergency management support.

For the Schelde Estuary an Evacuation Support System (ESS) was developed and applied in the flood prone area of Walcheren and Zuid-Beveland. The ESS supports decision makers in developing evacuation plans, by providing relevant information on the areas at risk of flooding. The ESS is a tool that links different breach locations to a database with simulations of flood events. The ESS also presents spatial information such as topographic data, the location of hospitals and postal code zones containing the number of inhabitants.

From the consultations with potential end users of the system it was concluded that the ESS provided stakeholders with an insight into the flood hazard and consequences. New flood event scenarios can be added to the system allowing an archive of flood events to be built up. In the future the management response module of the ESS is intended to be able to store several evacuation plans that can then be compared.

Keywords: Decision Support System, Flood event management, Schelde Estuary, Thames Estuary

Modelling tsunami overtopping of a sea defence by shallow-water Boussinesq, VOF and SPH methods

Peter Stansby, Rui Xu & Ben Rogers
School of Mechanical, Aerospace and Civil Engineering, University of Manchester, UK

Alison Hunt
School of Engineering, University of Plymouth, UK

Alistair Borthwick & Paul Taylor
Department of Engineering Science, University of Oxford, UK

Tsunamis cause devastating losses as they run up and overtop nearshore regions. While considerable basic experimental and numerical modeling has been undertaken for runup, little research appears to have been done on the more challenging problem of overtopping of embankments or sea defences. Here we consider overtopping of a trapezoidal structure with 1:2 side slopes for which there is accurate data, albeit for one case, from the Coastal Research Facility at HR Wallingford. The solitary wave (tsunami) is modelled in three ways using:

1. the Boussinesq/shallow-water model with empirical criteria for the onset of breaking, bore diffusion and bed friction (open source program WOT)
2. the Volume Of Fluid (VOF) method, through the commercial code STAR-CD, which is of more general form with turbulence modeled through k-ε or k-ω transport. In principle this requires no empirical input other than that embodied in the turbulence model.
3. the SPH (smoothed particle hydrodynamics) method, applying the open source SPHYSICS code, which is a Lagrangian particle method. This is ideally suited to violent surface motions but is traditionally quite dissipative for propagating waves, which is avoided here using a Riemann solver.

The methods are quite different in nature. The Boussinesq/shallow-water model is basically a 1-D model with three empirical parameters for the gross effects associated with depth-limited breaking. The VOF method is a 2-D vertical plane model based on the finite volume method with an interface capturing scheme and turbulence modelling. VOF methods have received massive commercial investment. SPH is a relatively new method for which no special surface treatment is required. It has shown high promise for violent flows such as dam breaks but here its suitability for inviscid wave propagation and overtopping is assessed. While versatile, it does require large numbers of particles and is computationally intensive. Representative computer times for each method on a single processor are respectively 1min, 20 hours, 50 hours. However the SPH method is ideal for parallel processing with high scalability. The geometry is shown below. For a wave of 0.1 m height in water 0.5 m deep the resulting volume overtopping the structure is 32 litres/metre. This is predicted within 10% by WOT but underestimated by VOF and SPH methods at present but higher resolution simulations are underway.

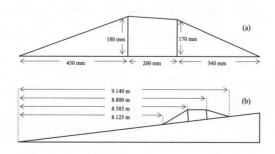

Comparisons with surface overtopping time histories will also be made. In addition some results will be shown for the wave group NewWave representing extremes in a spectrum; this actually contains three waves and there is thus influence of wave-wave interaction.

Keywords: Tsunami, wave overtopping models, inundation

Modelling the 2005 Carlisle flood event using LISFLOOD-FP and TRENT

J.C. Neal, P.D. Bates & T.J. Fewtrell
School of Geographical Sciences, University of Bristol, Bristol, UK

N.G. Wright
UNESCO-IHE Institute for Water Education, Delft, The Netherlands

I. Villanueva
OFITECA, Malaga, Spain

N.M. Hunter
JBA Consulting, Skipton, North Yorkshire, UK

M.S. Horritt
Halcrow Group Ltd., Swindon, Wiltshire, UK

Recently, the focus of flood inundation research has shifted from rural to more urbanised catchments. This change has been driven not only by economic and political factors, but also, by the increased availability of the computational power, digital elevation data and digital map data necessary to simulate spatially distributed flow patterns between buildings. To date, much of the flood inundation research in urban areas has concentrated on benchmarking model codes against one another, assessing the sensitivity of simulations to model parameters and structure, developing new modelling tools that integrate surface and sewer flows, and developing porosity methods that incorporate sub-grid topographic information. However, instances where the accuracy of an urban inundation simulation has been assessed against observations of inundation extent or water surface elevation are rare due to a lack of validation data. As such, little is known about the performance of flood inundation models over the complex and varied terrain found in urban areas relative to rural areas often dominated by pastoral land cover. Significant uncertainty remains over the most suitable choice of model code, parameters values and spatial resolution in urban areas, especially as this may vary with attributes of the urban fabric such as building separation and density.

This paper reports one of the most comprehensive validation data sets available on an urban fluvial flood, collected shortly after a 2005 event in the city of Carlisle, UK. For the first time, this data set provides 330 maximum water level and extent measurements in conjunction with gauged inflow hydrographs to define model boundary conditions, stage time series at three locations within the model domain, LiDAR elevation data and digital MasterMap™ data. This data set was used to assess the accuracy of two flood inundation models based on the LISFLOOD-FP and TRENT codes at various resolutions.

The accuracy of model simulations given a range of plausible roughness values was calculated in four areas of the city, which can be distinguished from each another by their building type, sizes and separations. A comparison of simulated and measured maximum water surface elevations indicate that root mean squared errors as small as 0.2 m could be obtained over open land and in some urban areas. However, blockage effects in other areas of the city, especially those typified by dense terraced housing, were found to prevent flood wave propagation and could lead to an increase in root mean squared errors, even when using a 10 m grid resolution. A long term aim of this research is to form part of a process that should ultimately lead to a set of recommendations on resolution requirements of raster diffusion wave models in urban catchments.

Keywords: Urban flooding, hydraulic modelling, model validation, Carlisle

Experience of 1D and 2D flood modelling in Australia – a guide to model selection based on channel and floodplain characteristics

J.M. Hannan & J. Kandasamy
University of Technology, Sydney, Australia

The average annual cost of flooding in Australia is $314 million per year (BTE, 2001). Some 100 flood studies, floodplain management studies and plans are currently being undertaken in Australia to seek to reduce the potential flood risk to residents and properties in flood-affected areas.

Consequently, a great body of knowledge and experience in flood modelling practices has been acquired, ranging from one-dimensional (1D) steady-state models through to two-dimensional (2D) finite element hydrodynamic models. Commonly used hydraulic modelling programs are: HEC-RAS (1D), MIKE-11 (1D), RMA-2 (2D finite element), TUFLOW (1D/2D finite difference), MIKE-FLOOD (1D/2D finite difference) and SOBEK (1D/2D finite difference).

It is recognised that there are a number of advantages in using 2D modelling. These include:

- the ability of the model to determine actual flow paths across the floodplain based on the terrain of the study area;
- simulation of the hydraulic interactions between the channel and floodplain, including momentum exchange, in the horizontal plane;
- water levels and flow velocities are available at every model grid point; and,
- the ability to employ fine model resolution near critical floodplain features.

However, 2D modelling has significant disadvantages when compared to 1D modelling, such as:

- potential loss of channel definition;
- more advanced data requirements;
- difficulty in data integration;
- greater expense both in terms of initial output and gathering data; and,
- longer computation times.

This paper critically appraises 1D and 2D hydraulic modelling techniques based on a quantitative comparison of the modelling results for two case studies in New South Wales, Australia. These case studies were selected to represent two different types of channel and floodplain systems with distinct principal hydraulic features. The hydraulic models for these case studies were developed using the MIKE-11, HEC-RAS and RMA-2 software packages.

For each case study, the reliability of the 1D and 2D modelling approaches is assessed to determine their applicability to each system. As a result, the paper provides practical guidance for modellers on the suitability of 1D and 2D modelling for physical channel and floodplain characteristics.

Keywords: Modelling, one-dimensional, two-dimensional, HEC-RAS, MIKE-11, RMA-2

Computationally efficient flood water level prediction (with uncertainty)

Keith Beven, Peter Young & David Leedal
Lancaster University, Lancaster, UK

Renata Romanowicz
Institute of Geophysics, Warsaw, Poland

Water level predictions made with hydraulic models are uncertain. Evaluating that uncertainty using Monte Carlo ensemble predictions is computationally expensive. While such ensemble methods have been used (albeit rarely) for probabilistic flood risk mapping and (even more rarely) for flood forecasting, it would be very useful, therefore, to be able to emulate the outputs of a fully distributed hydraulic model using a computationally efficient approach. The outputs from a distributed hydraulic model are, however, nonlinear with dynamics that depend on the nature of the inputs and the spatial patterns of flooding in a particular event. In this paper we show how a simple nonlinear transfer function approach can be used to reproduce with high accuracy the outputs of a 1-D hydraulic model (HEC-RAS) at every cross-section along the Montford to Buildwas reach of the River Severn, with estimates of the uncertainty in the predictions.

The methodology requires the identification of a nonlinear state dependent filter on the inputs + a linear transfer function to represent the reach dynamics. Levels are predicted directly, but can be mapped back to inundation extent using the cross-sectional information of the hydraulic model. The demonstration on the River Severn shows how this nonlinear transfer function model can be accurate over a wide range of upstream inflows. Where measurements of water level are available in real time, the methodology can also be used with a data assimilation algorithm. Where backwater or tidal forcing effects are important, the methodology can be used with both upstream and downstream inputs.

The methodology can also be extended to act as an 'emulator' in wider sense sense: namely emulating the response of the hydraulic model in the whole region of its defined parameter space. In this way, given some historical inundation extent data for the cross-sections it might be possible to identify the set of hydraulic parameters that will match the observations in a much more computationally efficient way.

Keywords: Flood inundation, Model Emulation, Data Assimilation, Severn, Uncertainty

Optimization of 2D flood models by semi-automated incorporation of flood diverting landscape elements

P. Vanderkimpen
Flanders Hydraulics Research & Soresma, Antwerp, Belgium

P. Peeters & K. Van der Biest
Flanders Hydraulics Research, Antwerp, Belgium

1 INTRODUCTION

Flood patterns can be affected by the presence of flood diverting linear landscape elements (dikes, road or railroad embankments) in the flood plain. When using a 2D regular grid flood model, the representation of these elements requires the use of a very fine elevation model grid. This often results in very large data sets and prohibitively long model run times. When elevation data are aggregated into a coarse grid, narrow linear landscape elements are frequently flattened or filtered out. Manually restoring these elements becomes impracticable when large areas are being studied. Therefore, a simple GIS-procedure was developed for restoring these elements in a semi-automated way.

2 PROCEDURE

Linear landscape elements were restored by combining raster (elevation) and vector (linear elements) data. First, elements that may affect flood patterns were selected from a vector data base. Subsequently, their crest levels were estimated from a high resolution raster elevation model. Next, a number of lower resolution raster elevation models were generated through aggregation of the high resolution elevation data. Finally, crest levels in the low resolution model were automatically restored by means of information previously extracted from the high resolution model.

3 APPLICATION

The procedure was applied to the eastern part of the Belgian coastal plain, which is characterized by the presence of a number of inland dikes, bordering a former tidal inlet.

Starting from a high resolution (5×5 m) DEM, four different elevation models were built: (1) raw fine (20×20 m), (2) raw coarse (40×40 m), (3) fine with linear elements restored and (4) coarse with linear elements restored.

The different elevation models were incorporated in a set of dynamically linked 1D-2D flood models, constructed by means of the software package MIKE Flood (DHI). These models were used to predict floods resulting from breaches in the coastal defence, caused by a series of extreme events (return periods 4000, 10000 and 40000 years). Afterwards, the hydraulic results from the different models (inundated area, water depth, flow velocity, ...) were compared.

In a final step, the results from the hydraulic models were used to estimate flood damage, number of casualties and risk. The values obtained with different elevation models were compared.

4 RESULTS

The results obtained with fine and coarse raw elevation models are quite different (flood risk differs by 30%), but when linear elements are restored, they agree remarkably well (flood risk differs by less than 1%). The coarse flood models, however, run four (and possibly eight) times faster than their fine counterparts.

5 CONCLUSION

When using a 2D regular grid flood model for flood risk evaluation, model performance can be improved considerably by means of a simple GIS-procedure for semi-automated incorporation of flood diverting landscape elements into the elevation model. Model run times can be reduced by an order of magnitude, while still preserving flood risk accuracy.

Keywords: Optimization, 2D flood models, landscape elements, flood risk

Understanding the runoff response of the Ourthe catchment using spatial and temporal characteristics of the storm field obtained by radar

P. Hazenberg, H. Leijnse & R. Uijlenhoet
Hydrology & Quantitative Water Management Group, Wageningen University, The Netherlands

L. Delobbe
Royal Meteorological Institute of Belgium, Belgium

In the year 2001 the Royal Meteorological Institute of Belgium (KMI) installed a C-band Doppler radar in the southern Ardennes region in Belgium near the border of Luxembourg at an elevation of 600 m. The 1597 km^2 Ourthe catchment lies within 50 km of this radar. As one of the bigger sub-catchments, this tributary forms an important input to the flood response of the river Meuse before it enters the Netherlands. So far, runoff response during extreme discharges have been difficult to simulate using only rain gauge data.

To assess the potential of the space-time structure of rainfall estimated using the radar for runoff modeling a collaboration between the Hydrological Service of the Walloon Region of Belgium (MET-SETHY), the KMI and Wageningen University was started. Ten rain gauges more or less equally distributed over the watershed measuring at an hourly interval are used to validate the radar. Near the outlet discharges are collected at the same time step.

A previous study (Berne et al., 2005) presented the first results of the radar application for six rainfall events. Basin average rainfall values were used as an input to the conceptual HBV model during a stratiform and convective event. Results showed that predicted discharges were highly dependent on initial conditions.

To diminish these effects, in this study the spatial and temporal characteristics of rainstorms and the resulting runoff response is being assessed over the period October 1, 2002 until March 31, 2003. During this winter half year storm events were mainly stratiform resulting in four important runoff peaks. First analyses show good correspondence between radar and rain gauge data. Most storm periods cover only a fraction of the basin. To quantify the hydrological relevance of the high spatial resolution of the radar as compared to the gauge data, both a lumped and distributed model are utilized.

REFERENCES

Berne, A., M. ten Heggeler, R. Uijlenhoet, L. Delobbe, Ph. Dierickx, and M. de Wit, 2005, A preliminary investigation of radar rainfall estimation in the Ardennes region and a first hydrological application for the Ourthe catchment. *Natural Hazards and Earth System Sciences*, 5, 267–274.

The importance of spill conceptualizations and head loss coefficients in a quasi two-dimensional approach for river inundation modelling

M.F. Villazón & P. Willems
Katholieke Universiteit Leuven—Hydraulics Section, Leuven, Belgium

In a quasi two-dimensional approach, floodplains are modelled by means of a one-dimensional hydrodynamic modelling system through a network of fictitious river branches linked to the main river with spills. The accuracy of the amount of water that passes through the spills to the flood branches is in such approach highly important. The choice of the most suitable model conceptualising the spills is, however, still a difficult task. This paper analyses the effect of different spill model conceptualisations as well as the effect of selected head loss coefficients in order to better understand the performance of quasi-2D floodplain models.

A reduced area (of 5 km length) of the river Dender in Belgium was implemented by means of three different river modelling softwares: Mike11 (DHI Water & Environment), InfoWorks-RS (Halcrow & HR Wallingford) and HEC-RAS (U.S. Army Corps of Engineers). The models were implemented, calibrated and validated maximizing the consistency between the models. Simulations were conducted for three historical flood events and eight synthetic events with return periods between 1 and 1000 years.

Evaluation and comparison was made of the results, for the maximum volume of water that passes through the banks to the floodplains, the peak discharges and the water levels in the floodplains.

First conclusion was that the three modelling systems in general show close results with a root mean square error of around 6 cm for the peak river levels. A similar error value was achieved after varying the spill's head loss coefficients. For the water levels in the floodplains, simulation results are different. Varying the spill's head loss coefficients shows variations less than 10 cm, wile the differences in results between the three modelling packages are as large as 25 cm after having maximized the consistency between the models.

Model conceptualisation thus appears more important in modelling the spill discharges than the selection of the head loss parameters is. Selection of the software's default values for the parameters leads in all cases to unbiased results after comparison with historical floodplain inundation data.

Keywords: Conceptualisation, floodplain modelling, sensitivity analysis, spills

Flood Risk Management: Research and Practice – Samuels et al. (eds)
© 2009 Taylor & Francis Group, London, ISBN 978-0-415-48507-4

Inundation scenario development for damage evaluation in polder areas

L.M. Bouwer & P. Bubeck
Institute for Environmental Studies, Vrije Universiteit Amsterdam, Amsterdam, The Netherlands

A.J. Wagtendonk
Spatial Information Laboratory, Vrije Universiteit Amsterdam, Amsterdam, The Netherlands

J.C.J.H. Aerts
Institute for Environmental Studies, Vrije Universiteit Amsterdam, Amsterdam, The Netherlands

Flat polder areas protected by dikes are particularly vulnerable to flooding, and are located in many delta areas in the world (Netherlands, Vietnam, USA, etc.). In these areas, flooding can occur because of dike breaches during high water levels. In the absence of historic flood loss data, the estimation of multiple potential loss events is needed to arrive at a realistic range of potential damages.

Two-dimensional hydrodynamic modelling has traditionally been used to create flood scenarios, from which potential losses can be calculated. However, hydrodynamic modelling is limited to a small number of scenarios that can be produced, and results in a relatively coarse horizontal resolution of the inundation depth scenarios. Due to the small amount of scenarios currently available, the uncertainties associated with the location of a potential dike breach and different quantities of inundating water is hardly taken into account in recent risk estimates.

We present an alternative approach, using a digital elevation model for creating a large number of flood scenarios in an area protected by one main dike. Within the study area along the river Meuse in the southeast Netherlands, a total of 22 individual sub-basins were identified. A set of flood scenario's have been created each consisting of different combinations of flooded sub-basins. Potential losses were calculated for simple depth-damage curves, that have been derived from the which were derived from the Dutch standard method for flood damage evaluations. The modelled flood losses are comparable to losses estimated in the Dutch project 'Flood risks and safety in the Netherlands' (Floris), for which hydrodynamic modelling was applied.

Our approach assumes that individual basins are filled up to the lowest part of the basin boundary each time. The approach ignores the fact that floods have a dynamic nature and that part of the losses are due to high flow velocities, assuming only losses from inundation. However, the advantage of our approach is that a large range of potential flood events can be created in relatively little time, and that it can be applied to very high-resolution elevation models, in our case 5 metres horizontal resolution. Both the large range of flood scenarios and the high horizontal resolution reduce at least part of the uncertainty encountered in flood loss modelling. Furthermore, this approach also allows taking numerous dike breach location and different quantities of inundating water into account. Using this simple approach, a better estimate can be potentially made of flood risks associated with large scale inundations in relatively flat polder areas.

Keywords: Flood modelling, economic loss, Netherlands

System analysis

Importance of river system behaviour in assessing flood risk

M.C.L.M. van Mierlo
Deltares (unit Inland Water Systems, until 2007 WL/Delft Hydraulics), Delft, The Netherlands

T. Schweckendiek
Deltares (unit Geo-engineering, until 2007 GeoDelft), Delft, The Netherlands
TU Delft, Section Hydraulic Engineering, Delft, The Netherlands

W.M.G. Courage
TNO Built Environment and Geosciences, Delft, The Netherlands

Extensive research on assessing flood risk has been carried out in the Netherlands in the last two decades. It is nowadays commonly accepted that a risk-safety based approach is indispensable in assessing the efficiency of flood protection strategies and measures. It was shown that effects of river system behaviour, especially the negative ones, can not be neglected in properly assessing flood risk in the Netherlands. A basic aspect of river system behaviour is that a local dike breach may affect hydraulic loads and hence dike failure probabilities at other locations. The flood risk (or safety) of a particular area may, therefore, depend on the safety of other areas. As a consequence, measures aiming at improving the safety of a particular area may have beneficial or adverse effects on the safety of other areas. Effects of river system behaviour are not yet considered in the current Dutch flood protection strategy.

In total two papers are submitted by the Delft Cluster project DC04-30, Safety from Flooding, Work package C. In this paper a framework for quantification of river system behaviour effects is presented. To this end, state-of-the-art modelling techniques of the hydrodynamic loads on the flood defences and the geotechnical strength as well as a module for the estimation of flood damage have been integrated into a probabilistic framework. The focus is on the description of the probabilistic framework and the implemented elements in abstract terms. Furthermore, the Importance Sampling scheme applied to reduce the number of computationally intensive hydrodynamic model calculations is described. The other paper on this project by van Mierlo, Courage and Schweckendiek will illustrate the application of the framework to a case study and give more details on the flood modelling.

Keywords: Flood risk, Failure mechanisms, Flood modelling, Flood damage, River System behaviour, Uncertainties

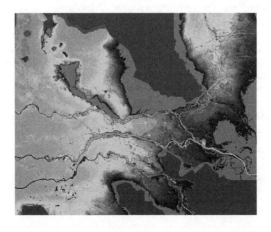

Development and evaluation of an integrated hydrological modelling tool for the Water Framework Directive and Floods Directive

M.B. Butts
DHI Water & Environment, Hørsholm, Denmark

E. Fontenot
DHI US, Morrisville, Philadelphia, USA

M. Cavalli, C.Y. Pin, T.S. Jensen & T. Clausen
DHI Water & Environment, Hørsholm, Denmark

A. Taylor
City of Boulder, Public Works, Boulder, Colorado, USA

Effective floodplain management at the catchment scale must address flood risk as part of a holistic water management strategy. From a European perspective water managers must address the complementary requirements of the Water Framework Directive and the Floods Directive. Integrating powerful flood mapping capabilities into a comprehensive hydrological modelling tool provides water managers with a valuable tool for both flood and catchment management that permits a more holistic approach to flood risk as a water management issue.

This paper presents the development and evaluation of such an integrated hydrological modelling tool. The hydrological component consists of the comprehensive process-based hydrological model, MIKE SHE. Within the MIKE SHE framework a coupled 1-D and 2-D hydraulic modelling capability has been developed to address the requirements of floodplain managers and the Floods Directive. In addition to supporting a more holistic approach, the resulting tools can address new problems such as groundwater generated flooding, surface water groundwater interactions particularly for wetlands, evaluating the importance of infiltration and evaporation effects on flood hazard mapping, and climate change assessments.

To evaluate this new tool, a systematic study has been carried out to demonstrate the value of 1-D and 2-D coupling. This paper presents the results for a case study in South Boulder Creek, Colorado, USA. Interest in assessing flood risk is motivated by the discovery previously unidentified flood prone areas, the subsequent substantial development that has occurred in this region without flood protection, and the need to improve the existing regulatory flood plain. A two-dimensional modelling approach combined with high resolution topography data is expected to simulate more accurately the detailed flow paths and properly capture the floodplain storage and therefore lead to more reliable flood maps. Therefore a systematic analysis of the effects of resolution in the topography is carried out. Traditionally, however, observation data is limited to flow or water level within the river channels and no observations for model validation are available in the floodplain areas at risk. Hydrological analyses showed the flood risk in this area was highly sensitive to groundwater levels and antecedent moisture conditions and therefore validation against other data was required. In this study model validation for a major historical flood is carried out using archived maps, public questionnaires and historical aerial photographs for model validation. The results show that the modelling approach, the data resolution and a proper hydraulic understanding of the floodplain flow paths are important for flood hazard mapping.

Keywords: Flood hazard mapping, flood risk, Water Framework Directive, Floods Directive, model credibility, stakeholder participation

A comparison of modelling methods for urban flood risk assessment

C.J. Digman, T. Bamford & D.J. Balmforth
MWH, Wakefield, UK

N.M. Hunter & S.G. Waller
JBA Consulting, Skipton, North Yorkshire, UK

Flood risk in urban areas is complex. It can arise from coastal inundation, river flooding, or local surface runoff. Often it can be a combination of all three events. Recent advances in modelling technology means that it is now possible to develop fully integrated and detailed models to assess such flood risk. However, these models can take a considerable amount of effort to build and verify, require substantial data, and are time consuming to run. As an alternative, simplified modelling techniques are often used, especially in arriving at a high level assessment of flood risk.

As a very broad rule, the more detailed a model, the more certainty there is in predicting the result, especially where the model is verified against field performance data. When using a more simplified and potentially unverified approach, the level of uncertainty needs to be quantified. The authors have undertaken trials on a number of catchment areas in the UK to determine the performance of different modelling approaches to assessing urban flood risk. These models vary from simple 2D surface models, to detailed models that fully integrate rivers and urban drainage systems, and below ground and above ground drainage.

The authors show what results can be expected from the different modelling approaches. They indicate where simplified models can be used with confidence and where they may provide misleading results. They indicate under what circumstances detailed modelling can be justified, and where cost/time savings can be realistically made by taking a simpler approach. The findings will be useful to both researchers and practitioners with an interest in urban flood risk management.

The authors have been involved with the UK Government's work on integrated modelling and integrated urban drainage pilot projects.

Keywords: Flood risk management, urban environment, modelling

Coastal flood risk analysis driven by climatic and coastal morphological modelling

M.J. Walkden, J.W. Hall, R. Dawson & N. Roche
Tyndall Centre for Climate Change Research, Newcastle University, Newcastle, UK

M. Dickson
Auckland University, Auckland, New Zealand

Effective shoreline management planning requires broad scale and long term coastal flood risk analysis. Such work has become more necessary, and challenging, as sea level rise and changes in other climatic variables have become more evident.

Coastal boundary conditions tend to complicate flood risk analysis. In particular the close coupling between flooding and shore morphology introduces considerable uncertainty, because beach level is a major determinant in the reliability of flood protection structures. Measures to prevent coastal erosion, such as cliff protection and groynes, are often in conflict with flood protection because they reduce the influx of sedimentary material to beaches, causing their denudation and concomitant reduction in the reliability of flood protection structures. Consequently future coast protection may increase the overall risk of coastal hazards and both flood and erosion risks must be accounted for in shoreline management planning. Sedimentary processes often involve large longshore distances, so that long lengths of coast have to be dealt with.

Climatic changes are expected to increase the rate of sea-level rise and change the characteristics of storm surge and wave climates. The best insights into future climates are being generated by global and regional scale climatic models. Employing the results of such models to drive coastal flood risk analysis is quite problematic. First, such models do not provide the required hydrodynamic conditions, and so their output must be propagated through (multi-scale) models of surge and waves. Secondly, such data is entirely synthetic and different to the data normally used during model calibration and validation. This has the potential to cause an artificial change in the behaviour of a flood risk model. Thirdly, the uncertainty in climatic model predictions is well recognised but poorly quantified. The computational expense of such modelling means that relatively few simulations can be used to explore uncertainty and typically short 'timeslices' are used to represent longer periods.

This paper will describe recent developments in an ongoing project focussed on quantification of the flood risk around the Norfolk Broads. The scales of the project are relatively large, the flood plain being around 340 km^2, and the analysis is conducted to 2100. Flood risk here is influenced by the morphology of around 40 km of heavily engineered eroding soft cliff on the neighbouring coast. The approach is one of Monte Carlo simulation, exploring system responses to probability distributions of loading conditions, and all possible defence structure behaviours. The flood risk model is one module of the Coastal Simulator, being developed by the Tyndall Centre for Climate Change Research. This includes a regional model of coastal evolution (and engineering) and a series of hydrodynamic models from regional to ocean scale, which are driven by the output of regional and global climate models. The outputs from this coupling are quantified to enable risks and adaptation options to be appraised using the *common currency* of risk.

This paper will focus on how the outputs of this broader system of models are used to drive the flood risk analysis. The climate scenarios are based on a pioneering ensemble of GCM, regional climate and storm surge model outputs, providing the basis for probabilistic interpretation of future surge and wave climates. An approach based on anomaly analysis is developed to moderate historic loading conditions to represent the future, and to bridge timeslices. The influence of future morphological changes on flood risk, driven by climatic changes effects on patterns of erosion and sediment transport, and coastal engineering, are incorporated through joint morphological/flood simulation.

Keywords: Coastal flood risk, climate change, coastal morphology

Micro-scale analysis of flood risk at the German Bight Coast

G. Kaiser, S.D. Hofmann & H. Sterr
Department of Geography, University of Kiel, Kiel, Germany

A. Kortenhaus
Lichtweiß-Institute for Hydraulic Engineering and Water Resources, TU Braunschweig, Braunschweig, Germany

Coastal lowlands in the North Sea Region are highly susceptible to be flooded in the course of extreme storm surges. A clear understanding of the likely risks and impacts of flooding is therefore compulsory. In the framework of the EU-project **FLOODsite** new approaches are developed for hazard analysis, vulnerability assessment, and risk management. In order to apply some of these new methodologies a pilot site application was conducted for the community of St. Peter-Ording at the German North-Sea Coast (Task 27 of **FLOODsite**).

Detailed risk analysis has been performed for this area. This analysis incorporated a probabilistic hazard analysis, a micro-scale vulnerability analysis, and the determination of flooding scenarios based on the hazard analysis. Finally, it has been developed into a GIS-based risk analysis tool merging the various categories of the economic, social and ecological vulnerability with scenario-based probabilities of flooding on a micro-scale level.

Within the hazard analysis, the annual probability of flooding of the hinterland was determined for each section of the flood defence system protecting the coastal area of St. Peter-Ording. This was achieved by detailing the hydraulic loading of the flood defences (sea dikes, dunes), including their uncertainties, and a detailed set of failure mechanisms with specific resistance parameters and uncertainties. Annual flooding probabilities were then determined by Monte-Carlo simulations of the Limit-State-Equations and a fault tree approach for each of the sections.

Based on the results of the hazard analysis for the flood defence system, the numerical SOBEK model was used to perform the flood inundation simulation for different flooding scenarios. A dike breach in the South of the area and heavy wave overtopping over a low asphalt dike section was agreed to be the standard scenario. Flood inundation simulations were then performed, also taking into account the uncertainties of the input parameters and different water levels during a storm surge. Results comprise inundation maps, water depths, and flow velocities.

In addition to hazard analysis, the estimation of the expected damages is crucial to assess the flood risk and its spatial distribution. This investigation aimed at improving two major deficits in vulnerability assessment, which are 1) the costs and efforts of detailed, object-based vulnerability studies and 2) the lack of social and ecological risk criteria.

Earlier vulnerability assessment studies have shown that only a micro-scale approach can deliver data, which are detailed enough to properly assess the flood risk, though it is rather time- and cost-intensive. Seeking for a minimisation of the effort for future damage potential analysis and damage estimations, a quick, economically feasible instrument for the precise evaluation of assets at risk and the damages has been developed.

In order to understand the interrelations of socio-economic and ecological dynamics as well as the intangible impacts of floods, an integrative methodology for vulnerability assessment has been applied. Following a multi-criteria risk assessment approach, the spatial distribution of economic, social, and ecological risk was investigated at the pilot site to identify specific risk zones. To assess the overall vulnerability, several risk criteria were assessed in St. Peter-Ording for each vulnerability category: (a) economic: buildings, private inventory, stock value, gross value added; (b) social: population at risk (& risk to life), vulnerable people, social hotspots, (c) ecological: coastal biotopes.

The multi-criteria risk assessment resulted in a comparative risk rating system for the economic, social and ecological damage categories constrained by different flooding scenarios. A GIS-based map output was considered the appropriate tool for a spatial analysis of the associated flood risk. The aforementioned results will be presented and discussed in greater detail during the conference and in the paper.

Keywords: Coastal flood risk, probabilistic hazard analysis, micro-scale vulnerability assessment

Flood hazard mapping for coastal storms in the Delta Ebro

D. Alvarado-Aguilar & J.A. Jiménez
Laboratori d'Enginyeria Marítima, ETSECCPB, Universitat Politécnica de Catalunya, Barcelona, Spain

Flood risk management is a critical issue in coastal low-lying areas. Flooding in coastal areas results from the combination of a driving agent, "the storm" normally defined in terms of a storm surge plus storm waves acting on a coastal stretch defined in terms of elevation. However, in coastal sedimentary environments, the impact of the storm will induce an impulsive and significant coastal response that will interact with the storm and should affect the intensity of the flooding (enhancing or reducing). Under these conditions, the manager must be aware of the storm-induced coastal response due to its intrinsic associated damage (coastal erosion hazard) and its potential synergic effect with flooding (increasing flooding hazard).

In this work, we present a methodology to take into account the intrinsic dynamic nature of coastal sedimentary areas in flood hazard mapping. It consists in incorporating morphodynamics into the hazard assessment to estimate its temporal variation during the flood event. To analyse the importance of this factor, three approaches (no beach response; known beach variability; simulation of storm-induced beach profile response) were applied to calculate the flooding in the Ebro delta induced by the largest recorded storm (November 2001) along the Catalan coast.

It has to be stressed that results of the overtopping rates are affected not only by the variability introduced by considering the response of a given stretch to the storm but also by the spatial variability in the response along the coast. Integrated water volumes have been calculated by estimating the beach freeboard (distance from the beach crest to the total water level) variation during the storm (due to changes in wave conditions and beach elevation, when applicable). These water volumes will mainly inundate the areas closest to the shoreline and constrained by levees or roads delineating the outer row of rice fields. The application of the different methods results in overtopping rates varying in one order of magnitude. When integrated during the storm duration this results in water volumes that can vary about 40%. Moreover when this is applied to different areas they are able to catch the spatial variability in the response to storms and associated inundation.

The used methodology combines the probabilistic analysis of water level (surge and waves), the coastal evolution modelling, flood modelling and mapping within a GIS framework.

Keywords: Flood hazard mapping, beach erosion, coastal storms, coastal hazards, Ebro delta

RAMWASS Decision Support System (DSS) for the risk assessment of water-sediment-soil systems – application of a DSS prototype to a test site in the lower part of the Elbe river valley, Germany

B. Koppe
LEUPHANA University of Lueneburg, Faculty of Environmental Sciences and Engineering, Suderburg, Germany

B. Llacay & G. Peffer
CIMNE, Gran Capità s/n, Barcelona, Spain

Global change and human activities can exert severe impacts on the ecology of aquatic and wetland ecosystems adjacent to a river basin area (hereafter termed fluvial ecosystems), influencing in some cases the life of many human beings. The devastating impacts of recent severe contamination accidents in relevant fluvial ecosystems (such as the Aznalcóllar mine dam failure and spill in the Doñana marshes adjacent to the Guadalquivir river estuary in Spain on 1998) have sparked interest in the water-sediment-soil modelling in fluvial ecosystems worldwide. Similar critical situations can be found in the preservation of the biosphere reserve of the Elbe river valley in Germany and the marshland and lagoons of the Po river delta in Italy.

Above examples, and many similar ones, demonstrate that there is a need for efficient methods and tools to assist public administrators and emergency services in the risk assessment of the water-sediment-soil (WASS) system in fluvial ecosystems and in the management of different prevention, mitigation and remediation scenarios.

We present the research being done within the EC-funded project RamWass, which started on 1st September 2006 and will finish on 28th February 2009. The objective of the RamWass project is to develop and validate a new decision support system (DSS) for the risk assessment and management for the prevention and/or reduction of the negative impacts caused by human activities on the water/sediment/soil system at river basin scale in fluvial ecosystems. The DSS integrates environmental and geo-physical data from earth observation systems, in-situ sensors, geo-referenced information, artificial intelligence tools, and advanced computer modelling, simulation and visualisation. The generated knowledge will contribute to the assessment of the ecological impact and the design of effective response actions maximising the integrity and safety of the ecosystem and human life.

A crucial activity of the RamWass project is the in-depth calibration, validation and assessment of the performance, scalability and effectiveness of the DSS in its application to three relevant fluvial ecosystems adjacent to important river basins in Europe:

1. The marsh area of Doñana in Spain
2. The biosphere reserve Elbe Riverland in the lower part of the Elbe river valley in Germany
3. The marshland and lagoons of the Po river delta in Italy.

This article will present the experience gained with the development of the RamWass DSS. It will be built in such a way that, after the adequate calibration, it can be effectively applied to the three testing scenarios chosen, as well as to any other fluvial ecosystems selected in the future.

Keywords: Water-sediment-soil, decision support system, risk management, fluvial ecosystems

Radar based nowcasting of rainfall events – analysis and assessment of a one-year continuum

H.-R. Verworn & S. Krämer
Institute of Water Resources Management, Leibniz University of Hannover, Hannover, Germany

Flash flood forecasting and warning as well as real time control decisions and activities require sufficient knowledge about the probable rainfall some time ahead. With their high resolution in space and time weather radar data represent the variability of the actual rainfall processes quite well. Rainfall structures within each radar data set can be identified dependent on intensity and size and may be re-recognized in following data sets. By analysing the spatial offset between the data sets, direction and velocity of the structures can be derived. The extrapolation of the cell structures using the movement vectors enables the estimation of future rainfall up to 2 hours ahead. This is generally understood as nowcasting.

There is, however, the problem that the cell structures do generally not keep their form and intensity distribution which is assumed in many cell tracking models. Instead, the structures may change more or less dependent on the event characteristics (stratiform or convective) resulting in increasing differences between nowcasted and true rainfall. This variable structure stability reduces the nowcasting potential and reliability.

To assess the quality and reliability of simple nowcasting with a cell tracking model, areal rainfall for a 51 km² catchment in the Emscher Area in Northrhine-Westfalia in Germany was nowcasted using the HyRaTrac model (Hydrological Radar Tracking). The radar data continuum of the Essen C-band radar (operated by DWD; 5 min, 16 classes, 1 × 1 km) from 2003 was used to extract all rainfall events to be used as input for HyRaTrac.

The mean squared error (MSE) between nowcasting and observation was calculated event-wise for various forecast horizons. By applying an error component analysis, the MSE can be divided into a linear and a non-linear portion which enables to derive information about the variability of the rainfall events and their nowcasting potential for operational applications. The results are systematically displayed and discussed dependent on event characteristics and forecast horizons.

Keywords: Radar rainfall, nowcasting

On the quality of Pareto calibration solutions of conceptual rainfall-runoff models

A.-R. Nazemi & A.H. Chan
Department of Civil Engineering, The University of Birmingham, Birmingham, UK

A. Pryke & X. Yao
School of Computer Science, The University of Birmingham, Birmingham, UK

Conceptual rainfall-runoff models are the most promising way to describe the hydrological behaviors in natural catchments. Recent investigations in calibration of conceptual rainfall-runoff models have suggested that it is almost impossible to assign a unique parametric vector which could perform well in all segments of the hydrograph. Therefore, it is more intuitive appealing to apply multi-objective search methods in order to calibrate conceptual rainfall-runoff models, in a way that similarity between the observed and simulated values in each segment of the hydrograph could be formulated as a single objective function. Applying multi-objective search methods will result to a set of Pareto optimal solutions which could be quite diverse in both parametric and objective spaces. In current paper, first, the essence and importance of visualization for multi-objective search solutions will be illustrated. After this brief introduction, three techniques will be demonstrated for visualizing the solutions related to multi-objective calibration of rainfall-runoff models. Then a novel visualization technique for non-dominated solutions, i.e., heatmap will be introduced which can contain all the information that the first three techniques can provide. Different applications of heatmap visualization will be explained briefly for the multi-objective calibration results of a conceptual model in Leaf River Basin. The introduced visualization techniques, afterward, will be applied on the multi-objective calibration solutions of two different models in MOPEX catchments and the results will be evaluated from different perspectives. Finally, the paper will be summarized and concluded.

Keywords: Conceptual rainfall-runoff models, multi-objective calibration, non-dominated solutions, visualization

Model reuse and management in flood risk modelling

Rahman Khatibi
Cascade Consulting (Environment and Planning) Ltd., Swindon, UK

Since the becoming of flood modelling as a business sector, model reuse has become widespread. Models are often commissioned by regulatory bodies or water undertakers which are consequently issued under licence for use by third parties. The focus of this paper is to identify and discuss issues on model management that are yet to be widely acknowledged. The term model is used here in the sense of datasets, serving inputs to modelling software application to produce outputs often for aiding decisionmaking. In this sense, it is easy to see the diversification of models and their impact on model reuse.

It is argued that modelling has reacted to opportunities and models have diversified. In flood modelling diversification is driven by:

- Different versions of the same model, e.g. model runs for a range of recurrence interval, climate change scenarios, various development scenarios
- Different users of the same model, e.g. regulating authorities issue models for third party users
- Application fields, e.g. design floods, real-time flood forecasting, maintenance scheduling
- Application problems e.g. low flows, high flows, water quality, sediment transport
- Software systems e.g. ISIS, MIKE11, Delft FEWS, HECRAS

Without transparency in modelling practices, diversification of models and modelling practices would be concerning, particularly in relation to model reuse. Some progress has been made through the emergence of modelling procedures, formulation of best practice guidelines and benchmarking of mathematical modelling software applications. In spite of these endeavours, the scope for variation in mathematical modelling is widening and many of the issues are often unresolved. Therefore concerns may be expressed on modelling in relation to defensibility, audit trail, integrity, transparency and model reuse. Since models are produced after significant investments, they should be regarded as assets; and since they often play key role in decisionmaking, they should be regarded as critical assets. This paper therefore highlights some of the model management issues and possible solutions.

International programmes

Flood Risk from Extreme Events (FREE): A NERC-directed research programme – understanding the science of flooding

Chris. G. Collier
Centre for Environmental Systems Research, University of Salford, Salford, UK

In carrying out research within the Earth system of specific relevance to flooding it will be necessary to,

- improve the estimation and prediction of flood risk from extreme events through considering the processes involved as an integrated system;
- seek ways to reduce uncertainty and improve the quantification of flood risk; and
- identify and articulate critical guidance on how flood risk is changing.

FREE is a £6 million, five year Natural Environment Research Council (NERC) directed research programme. It comprises thirteen individual projects, led by academic institutions managed by a Steering Committee and a NERC appointed Science Coordinator supported by NERC staff. These projects are mainly consortia, and involve a number of stakeholders including the Met Office, Defra and the Environment Agency. Overall the specific objectives of the programme are,

Objective 1: To develop and extend the science underpinning integrated modelling frameworks enabling models to work sensibly and more effectively together.

Objective 2: To identify and spread scientific improvements in model initialisation, data assimilation and the processing of forecast ensemble outputs across modelling communities.

Objective 3: To understand and quantify the propagation of uncertainty within a changing environment and within rapidly changing catchments.

Objective 4: To develop techniques for uncertainty reduction at the output stage enabling the communication of risk assessment to the user community.

Objective 5: To enable mitigation of, and adaptation to, floods by the provision of advice on flood risk management system development.

Objective 6: To seek the use, or re-analysis and use, of existing datasets whilst organising and creating a methodology to enable specific data to be collected following any extreme event that might occur during the period of FREE.

Objective 7: To engage with the national and international community and stimulate knowledge transfer and user engagement.

Objective 8: To provide training opportunities through research studentships including CASE awards, and to encourage the involvement of students funded via other means in FREE activities.

In this paper we will summaries the work on-going in FREE and highlight some of the initial outputs.

Advances in flood risk management from the FLOODsite project

P.G. Samuels, M.W. Morris & P. Sayers
HR Wallingford, Wallingford, Oxfordshire, UK

J-D. Creutin
LTHE Grenoble Cedex 09, France

A. Kortenhaus
Technische Universität Braunschweig, Braunschweig, Germany

F. Klijn, E. Mosselman & A. van Os
Deltares, Delft, The Netherlands

J. Schanze
Leibniz-Institut für ökologische Raumentwicklung e.V., Dresden, Germany

FLOODsite is the largest ever EC research project on flood risk management, with an EC grant to the budget of nearly €10 Million complemented by supporting national funds. The project, which started in 2004, is scheduled to complete in February 2009, and has involved over 200 researchers from 13 countries including many of Europe's leading institutes and universities. The project is interdisciplinary integrating expertise from across the physical, environmental and social sciences, as well as spatial planning and management. FLOODsite is an ambitious project to maintain the world-leading leading position of Europe in knowledge and practice for flood risk management. The pilot studies have drawn together the development testing of the project knowledge and provided feedback from flood risk managers and river, estuary and coastal stakeholders. The use of the pilot sites and collaboration with executive agencies in several countries should ensure that FLOODsite results are of real value, practicable and usable.

The circumstances in which the research will be implemented are changing internationally with development of policy through the entry into force in November 2007 of the European Directive on the assessment and management of flood risks and the concern of the potential for increased flood hazards arising from climate change as set out in the IPCC Fourth Assessment Report, published in November 2007. The scope of flood risk management is broad with FLOODsite providing incremental contributions to knowledge in several areas. FLOODsite has examined several aspects of flood risk management from appropriate policy and instruments, analysis methods, event management and decision support. The research has concentrated on specific topics identified in response to the original call for research as it is set in a much broader context of national and international research projects. During the research the project team has identified links to over 80 other projects and programmes.

The management, mitigation and reduction of future flood risk will not come from a single technical solution or policy but from a portfolio of responses which are tuned to the specific circumstances at a local or regional scale, taking account of national governance structures and public attitudes towards flood risks. This diversity of approach is recognised by the embodiment of the subsidiarity principle in the European Directive. This paper will cover the main areas of innovation achieved within FLOODsite and show how they may facilitate the implementation of the European Directive through flood risk assessments, risk mapping and the preparation of flood risk management plans. The project does not propose a single integrated methodology for flood risk management; rather it provides a set of linked methodologies which support integrated flood risk management. Other papers at this conference will present more detail on specific tasks and themes of the research than is possible in this overview.

For more information on FLOODsite visit www.floodsite.net.

Keywords: Flood risk management, research and development

The Tyndall Centre Coastal Simulator and Interface (CoastS)

R.J. Nicholls, M. Mokrech & S.E. Hanson
Tyndall Centre for Climate Change Research, University of Southampton, Southampton, UK

P. Stansby & N. Chini
Tyndall Centre for Climate Change Research, Manchester University, Manchester, UK

M. Walkden, R. Dawson, N. Roche & J.W. Hall
Tyndall Centre for Climate Change Research, Newcastle University, Newcastle, UK

S.A. Nicholson-Cole, A.R. Watkinson & S.R. Jude
Tyndall Centre for Climate Change Research, University of East Anglia, Norwich, UK

J.A. Lowe
Met Office Hadley Centre (Reading Unit), Reading, UK

J. Leake & J. Wolf
Proudman Oceanographic Laboratory, Liverpool, UK

C. Fontaine, M. Rounsvell & L. Acosta-Michlik
University of Edinburgh, Edinburgh, UK

The threat of sea-level rise and climate change, among other factors, means that coastal engineers and planners are being increasingly asked to make long-term assessments of potential coastal impacts and responses, especially for flooding. This is a challenging task as the coast is an integrated system, and interventions in one sector may influence the impacts for another sector. Given the multiple requirements of any intervention, an integrated assessment across a wide range of issues is required to make sound engineering decisions. In response to this challenge, the Tyndall Centre for Climate Change Research is developing a Coastal Simulator and Interface which integrates a wide range of coastal engineering and related models in way that supports long-term assessment and decision-making. This paper presents an overview of this effort, including the flooding components, while the companion abstract submitted by Walkden et al considers the details flooding under probabilistic beach conditions, and wave/ surge climates derived from Climatic General Circulation Models.

The prototype Coastal Simulator is based on a series of linked models within a nested framework which recognises three spatial scales: (1) the global scale; (2) the regional scale (in this case the North Sea and East Anglia); and (3) the simulator domain (a physiographic unit such as a coastal sub-cell). Within the nesting, the larger scale provides the boundary conditions for the smaller scale. The models feed into each other and describe a range of relevant processes, such as flood risk and how flood risk is influenced by changes other than climate change. Communication of the results is a major issue and the simulator includes a dedicated GIS-based user interface which presently allows a wide range of queries of libraries of model outputs, making the results available to a range of users. In addition to standard maps, three-dimensional visualisation of the results have been developed and have been found to be very effective in communicating the model results, especially to non-technical audiences. As the simulator interface is developed, some of the simpler models (e.g. the changing built environment) will be added to it to allow more user control of queries. Throughout the project, representation of uncertainty is a key issue, which is being addressed in a variety of ways, including driving the models with probabilistic climate change inputs to develop probabilistic outputs, where possible.

The pilot simulator is for sub-cell 3b where there are estimated to be about 1,400 properties within 100 metres of the cliff-top and 20,000 properties in the coastal flood plain, including 3,000 non-residential properties. The Simulator considers the climate drivers (waves, surges and mean sea level), wave transformation, shoreline morphodynamics, ecosystem change, and erosion and flood risk, including algorithm-based scenarios of the built environment. Shoreline management scenarios are developed with relevant stakeholders, and the options to be assessed are currently being extended. Presently, the simulator contains a library of 180 discrete simulations from the first phase of the Tyndall research, based on nine different climate scenarios describing sea level and wave conditions, five different shoreline management scenarios, and four socio-economic scenarios of the built environment. In terms of climate change, sea-level rise is the most important driver of change, with changes in wave conditions causing secondary effects. Shoreline management decisions also have a profound effect on downdrift sediment supply and the analysis suggests that removing cliff defences may greatly reduce flood in downdrift locations. Hence, for the first time these prototype Simulator quantify the possible trade-offs in erosion and flood risk that have been widely discussed in qualitative terms. The issues raised in developing the Simulator are generic to improving long-term coastal assessments and engineering interventions and the simulator concept should find wide application.

Keywords: Climate change, sea-level rise, flooding, erosion, integrated assessment

The social impacts of flooding in Scotland: A national and local analysis

A. Werritty, D.M. Houston, M. Jobe, T. Ball, A.C.W. Tavendale & A.R. Black
School of Social and Environmental Sciences, University of Dundee, Dundee, Scotland, UK

A key policy objective of the Scottish Government (as it transposes the new EC Floods Directive into Scots law) is to promote sustainable flood management, a key component being to promote "the maximum possible social and economic resilience against flooding". However, the social impacts of flooding across Scotland are very poorly understood at present. This paper reports on a project which begins to fill that gap by assessing the range of impacts that recent floods have had on people, their attitudes and their behaviour in living with flood risk. A household survey in Brechin, Edinburgh, Elgin (flooded twice), Forres, Glasgow, Hawick, Perth and scattered rural and coastal communities undertaken in 2006 yielded returns from 633 households which had been flooded over the period 1993–2005 and 590 returns from households not flooded in the same period. The results have been analysed both nationally (aggregating across all surveyed communities) and locally (reporting findings for each of the eight communities).

At the national scale, immediate intangible impacts (eg the stress of the flood itself and disruption from having to leave home) registered most severely on the sampled population, followed by lasting intangible impacts (eg worry about future flooding and loss of irreplaceable items) with tangible impacts (eg direct financial loss and decline in house values) coming last. All three groups of impacts were felt more severely by vulnerable populations (households with elderly or disabled members, low incomes and modest educational qualifications). Whereas buildings insurance was purchased by 90% of all households (with average losses of $c.$ £32,000), contents insurance only covered 70% of households (average loss $c.$ £13,500). Flood warnings, mainly via neighbours and council officials, had been received by around two-thirds of those flooded. Since being introduced by SEPA in 2002, *Floodline* has been highly valued, but this has yet to enhance generally low levels of flood awareness. Amongst those flooded, worry about future flooding is high and only two thirds of households were confident in what to do in future floods. Structural defences (especially walls and barriers) remain the most favoured measures for mitigating flood impacts, with local authorities deemed to be primarily responsible for their provision. However, some degree of personal responsibility was reported by 25% of respondents and 60% of households at risk had some form of flood-proofing.

At the local scale, there is a marked variation in flood impacts reflecting flood history (how severe, how recent and how often) and the socio-economic profile of the flooded communities (proportions of disabled and elderly, income and educational levels). Glasgow, Elgin and Perth typically reflected more severe impacts (notably in terms of length of temporary accommodation, stress of the flood, financial loss, worry about future floods, and loss of sentimental items). This implies that socio-economic resilience to flooding is spatially differentiated across Scotland with, by implication, contrasting levels of social capital available to mitigate flood impacts. The Scottish Government's aspiration to promote "the maximum possible social and economic resilience against flooding" and the development of flood management plans required under the new EC Directive, will require mapping of the most vulnerable communities and better characterisation of flood risk (especially in urban areas where the risk of pluvial flooding is high). A major paradigm shift is emerging as flood risk managers shift from defending vulnerable places via structural defences, to protecting vulnerable people, increasingly via non-structural measures.

Keywords: Social impacts of flooding, socio-economic resilience, vulnerable communities, spatial differentiation

The Flood Risk Management Research Consortium (FRMRC)

I.D. Cluckie
*Water and Environmental Management Research Centre (WEMRC), Department of Civil Engineering,
University of Bristol, Bristol, UK
Chairman of the Flood Risk Management Research Consortium, UK*

The flood Risk Management Research Consortium (FRMRC) was established under the auspices of the United Kingdoms Engineering and Physical Sciences (EPSRC) Research Council in 2004 along with a wide group of funder's from industry, research sponsorship and public service. It effectively represented a unique attempt to construct a consortium of researchers from across the UK not unlike an integrated programme (IP) generated by the European Union, similar to FLOODsite. This consortium was funded in the first instance for four years and this paper briefly reviews some of the key outputs that are now starting to appear. In the latter stages of 2007 the second phase of this program was initiated and is presently being implemented. This paper sets out to describe some of the successes, failures and developments to this strategic programme in relation to flood risk management research. The second phase has widened its region of support to include Eire and the development of key strategic international relationships. The current plans for the research focus of 'FRMRC2' will be outlined in the context of the themes of FLOODrisk2008.

Keywords: Flood Risk Management

EIB financing for flood risk mitigation

Christoph Gleitsmann
European Investment Bank, Projects Directorate, Luxembourg

1 EIB LENDING FOR NATURAL DISASTER IMPACT MITIGATION

In recent years, the EIB has seen an increase in lending for reconstruction and rehabilitation projects following natural disasters, for example floods, forest fires and earthquakes. This has been the result of (i) the growing frequency of these events, partly a possible indication for climate change, (ii) the willingness of public authorities to fund the reconstruction process through loans rather than grants, and (iii) the Bank offering special loan conditions for rebuilding after catastrophic events within the EU and neighbouring countries.

While EIB loans have concentrated traditionally on the reconstruction of damaged infrastructure and short-term clean-up operations, a number of projects financed by the Bank have focused on preventive measures, in particular flood risk mitigation, the main purpose being to safeguard the urban environment. In this instance, decision-making is essentially based on the economics of damage prevention combined with the environmental impact of the defences themselves.

2 CASE STUDY

2.1 *Support to the national flood prevention strategy, Czech Republic*

In the last decade, Europe has been hit hard by more than 100 major floods that have killed 700 people, displaced half a million others and caused at least EUR 25 bn in economic losses. The Czech Republic suffered 97 casualties and EUR 4.7 bn in flood damages, thus topping the list of affected EU-27 countries in terms of the cost of flood damages as a percentage of gross national product. Floods are natural phenomena that cannot be fully prevented; however, risks associated with flooding can be mitigated. The EIB has been funding extensive programmes in the Czech Republic and other affected EU countries for reconstruction and rehabilitation of infrastructure in the aftermath of major floods. It is complementing these efforts with increasing support for the preparation and implementation of flood prevention measures. The Czech Republic approved its National Strategy for Protection against Floods in the year 2000. The implementation has been split into two phases, both of which are supervised by the Ministry of Agriculture and are receiving total co-funding of EUR 360 m from the EIB. The first phase (2002–07) concerns more than 400 schemes protecting against floods in particular major conurbations. The second phase (2007–12) includes a wide range of further measures to mitigate flood risks in all Czech river basins and complement the increased protection of a total of 850 000 people. The project is compliant with the objectives of the EU Water Framework Directive as well as with the new Directive on Flood Risk Mitigation. Flood prevention measures are often environmentally sensitive and can have a significant impact on Natura 2000 areas. At the Bank's request, an independent environmental expert is assessing the soundness and quality of the environmental analysis and also provides support to the implementing entities, and independent strategic experts analyse the economical efficiency of proposed measures, to be in line with the national strategy.

This project is serving as a template for similar projects that are and will be prepared within the EU and beyond. The EIB is in the process of publishing a "Guide for preparation of flood risk management schemes", prepared by a specialised Consultant that should assist Governments and other stakeholders in these efforts. In the context of its new water sector lending policy, the EIB recognises the need for supporting all efforts to adapt to climate change within Europe and beyond.

Keywords: European Investment Bank, flood risk mitigation, Czech National Strategy for Flood Prevention

One nation, one policy, one program flood risk management

P.D. Rabbon, L.J. Zepp & J.R. Olsen
U.S. Army Corps of Engineers, USA

1 BACKGROUND

The US Army Corps of Engineers (USACE) and the Federal Emergency Management Agency (FEMA) are the lead Federal agencies in the area of flood damage reduction and floodplain management. The Corps' programs primary focus is on management of flood water (such as levees) and management of floodplains (such as elevating structures and floodplain acquisitions) to reduce the flood risks. FEMA programs emphasize the use of hazard mitigation programs, floodplain management measures and flood insurance to mitigate flood related losses.

State and local agencies are involved with flood risk management thru partnerships with the Corps and FEMA. They also have vital flood risk management responsibilities that are separate from the federal government, such as establishing building code requirements and regulating land use.

Several key connections, or "nexus", exist between the two Federal agencies' programs and the nonfederal governments. For example, under the FEMA administered National Flood Insurance Program (NFIP), all areas in a community designated as a Special Flood Hazard Area (SFHA) are subject to mandatory flood insurance purchase requirements and development restrictions. The condition and performance of USACE constructed levees play a role in determining what areas of a community are subject to flood insurance. The local community has a significant role as it is their safety, quality of life, and economic stability that are at risk.

Not only do programs and policies need better coordination to improve program results, but inconsistencies should also be addressed. For example:

- Nonfederal BVCost-Sharing: For Corps flood damage reduction projects authorized after 1996, nonfederal local sponsors must pay 35% of the total project costs. For FEMA mitigation projects, the local sponsor pays 25% of costs.
- Benefit-Cost Analysis: The Corps must follow detailed procedures for benefit-cost analysis as described in the Economic and Environmental Principles and Guidelines for Water and Related Land Resources Implementation Studies, while FEMA follows the less stringent Circular No. A-20 issued by the Office of Management and Budget.
- Levee Certification Standards: FEMA requires three feet of freeboard above the 1-percent-chance flood to certify levees for purposes of the NFIP, while the Corps uses risk-based procedures. The risk-based approach could result in either a higher or lower levee height depending on the uncertainty of the situation.
- Deed Restrictions: If a local government purchases property using FEMA mitigation funds, required FEMA deed restrictions could prohibit future USACE flood damage reduction projects.

Flood risk management policies and programs should be considered on a cross-agency compatible basis rather than an organizational basis. Solutions would be more comprehensive and avoid the "that's not our responsibility" syndrome.

2 COLLABORATIVE FRAMEWORK

The mission of USACE's National Flood Risk Management Program (NFRMP) is to integrate and synchronize the ongoing, diverse flood risk management projects, programs and authorities of USACE with counterpart projects, programs and authorities of FEMA, other federal agencies, state organizations and regional and local agencies.

Over the past two years, the Corps and FEMA have implemented an initiative called "Silver Jackets". The intent is to develop federal-state teams to better coordinate existing agencies authorities in flood risk management. The State of Ohio pilot has shown promise in fulfilling the goals of the initiative and additional teams are forming in other states.

The General Accountability Office released a report entitled, "NaturalHazard Mitigation: Various Mitigation Efforts Exist, but Federal Efforts Do Not Provide a Comprehensive Strategic Framework," GAO-07-403, August 22, 2007. This report concludes that FEMA, other federal agencies, and nonfederal stakeholders have collaborated on natural hazard mitigation, but the current approach is fragmented and does not provide a comprehensive national strategic framework for mitigation. Collaboration typically occurs on a hazard-specific basis, after a disaster, or through informal methods. A comprehensive framework would help define common national goals, establish joint strategies, leverage resources, and assign responsibilities among stakeholders.

3 FLOODPLAIN MANAGEMENT 2050

By 2050, we should have a unified approach to flood risk management between all levels of government. We should evolve into a single national vision and policy for flood risk management. The policy should be applicable vertically and horizontally within the federal, state, and local governments. The vision and policy should be based on a shared responsibility through partnering and collaboration to achieve long term economic, environmental, and socially sustainable flood risk management, which improves public safety and reduces flood risk through a comprehensive watershed approach.

Toward a transnational perspective on flood-related research in Europe – experiences from the CRUE ERA-Net

A. Pichler
Federal Ministry of Agriculture, Forestry, Environment and Water Management, Vienna, Austria

V. Jackson
Environment Agency, Wallingford, UK

S. Catovsky
Department for Environment, Food and Rural Affairs, London, UK

T. Deppe
Project Agency for Water Technology and Waste Management, Dresden, Germany

The management of flood risk is a critical component of public safety and quality of life. In the past, EU member states and associated states have mainly promoted their own national flood research without comprehensive co-ordination between their programmes, leading to duplication of effort and little opportunity to identify potential synergies. The flooding ERA-Net, CRUE, has accomplished a number of vital exchange and strategic activities that build the basis for the creation and implementation of a European Research Area (ERA) in flood risk management. The experiences made in these activities will contribute to a better understanding of flood risk and will help establish an appropriate evidence base to support policy-makers and key decision makers in terms of achieving future flood-resilient communities in Europe.

Infrastructure and assets

Flood Risk Management: Research and Practice – Samuels et al. (eds)
© 2009 Taylor & Francis Group, London, ISBN 978-0-415-48507-4

Hazards from wave overtopping

William Allsop
Coastal Structures, HR Wallingford, Wallingford & University of Southampton, UK

Tom Bruce
Division of Engineering, Edinburgh University, Edinburgh, UK

Tim Pullen
Coastal Structures, HR Wallingford, Wallingford, UK

Jentsje van der Meer
Van der Meer Consulting, Heere nveen, The Netherlands

Many flooding studies concentrate primarily on flood extent, flood depths and/or flood volumes. This narrow focus may be appropriate for diffuse flooding or for sites detached from the primary flood sources, but for flood risks from wave overtopping or dam break, processes close to the flood source may dominate hazards and risks. This paper will note recent advances in the methods to predict wave overtopping (abstract by Pullen *et al*); on present guidance on the levels of direct hazards caused by wave overtopping; and will discuss key areas where present knowledge and guidance are insufficient.

Recent work in UK, Netherlands and Germany has improved guidance on predicting overtopping in the new EurOtop Manual http://www.overtopping-manual.com/, drawing together results of national and European research projects and including the best available prediction methods and advice. Those tools are based primarily on field and laboratory studies, most of which require simplifying the structure being analysed. This paper will identify more complex methods to identify hazards from overtopping, and will highlight key research areas to improve future hazard prediction tools.

Emphasis on mitigating flood risks, adaptation of defences, and improving resilience of infrastructure, require better guidance for managing coastal defences. Reducing flood risks requires methods to quantify and reduce hazards to individuals or property, perhaps by warning individuals against overtopping hazards, particularly as studies under CLASH (see: http://www.clash-eu.org/) suggest that on average 4-8 people are killed each year in UK (and similar in Italy) through direct effects of waves on seawalls and similar structures. It is difficult to find records of overtopping damage to buildings or transport operations, but CLASH and site specific model tests (abstract by Udale-Clark *et al*) suggest that overtopping loads may easily exceed those against which typical building facades are designed. Stoppages of trains on coastal railway lines under wave overtopping confirm that these hazards can be significant, even under frequent conditions. So future guidance will require improvements on:

a. What overtopping can be tolerated by grass embankment or standard promenade protection?
b. Safe velocities/depths of overtopping (intermittent flows) for pedestrians or vehicles?

c. Violent (impulsive) overtopping tolerable by pedestrians or vehicles?
d. How do the effects of overtopping alter with distance and/or wind speed?
e. Loadings on buildings or secondary defences subject to wave splash or spray overtopping?
f. How can impulsive loadings from wave splash or spray be scaled to full scale?

The presentation will illustrate use of example methods to predict overtopping; will highlight current guidance on direct overtopping hazards to people and to property; and will discuss key gaps in current knowledge. It will be completed by suggesting methods to appropriately define and represent overtopping hazards in flood risk assessments.

Keywords: Flood risk assessment, wave overtopping, wave loading, safety analysis

Time-dependent reliability analysis of anchored sheet pile walls

F.A. Buijs & P.B. Sayers
HR Wallingford, Wallingford, UK

J.W. Hall
Newcastle University, Newcastle-Upon-Tyne, UK

P.H.A.J.M. van Gelder
Delft University of Technology, Delft, The Netherlands

Quantitative risk and reliability methods provide a rational basis to the design and operational management of flood defence systems. Flood defence reliability analysis represents failure mechanisms of the flood defence in the form of limit state equations. In these equations one or more variables can be subject to time-dependency. This paper illustrates how to capture the long-term time-dependent behaviour of flood defences in a reliability-based approach. A modelling methodology for statistical models of time-dependent processes of flood defences is presented and applied to anchored sheet pile walls (figure 1).

The first phase in the modelling methodology is problem formulation. This phase specifies the time-dependent variables of interest and the requirements of the statistical model. The second conceptualisation phase consists of a five-step analysis of the time-dependent process. Firstly, existing field observations and scientific understanding about the time-dependent process are assembled. Secondly, the excitation, ancillary and affected features and uncertainty types of the time-dependent process are analysed. Excitation features are flood defence properties that drive the time-dependent process, e.g. wave climate. Ancillary features are flood defence properties that contribute otherwise to the time-dependent process. Affected features are the flood defence properties subject to time-dependency. The third step describes the character of the time-dependent process. For example, renewal, cyclical, linear, logarithmic or history-dependent behaviour. The fourth step analyses the dependencies between different time-dependent processes. The fifth step formulates alternative statistical models for the time-dependent process. The last phase in the modelling methodology is parameter estimation, calibration

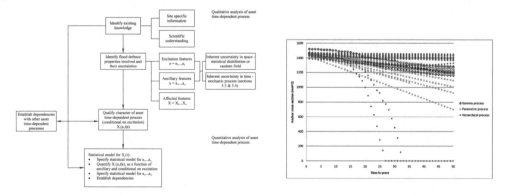

Figure 1. Left: modelling methodology for time-dependent processes. Right: example of time-dependent model for anchor corrosion.

and model corroboration. Completion of this phase often poses a problem as the availability of observations on time-dependent processes is limited.

The method is applied to the anchored sheet pile walls along the Dartford Creek to Swanscombe Marshes flood defence system along the Thames Estuary. The processes corrosion of the sheet pile wall surface, anchor corrosion and toe accretion or erosion are modelled. These processes are then embedded within a time-dependent reliability analysis. Time-dependent fragility and lifetime probability are calculated.

Keywords: Flood defence reliability analysis, deterioration, time-dependent processes

Analysis of tsunami hazards by modelling tsunami wave effects

Tiziana Rossetto
University College London, Civil, Environmental & Geomatic Engineering, London, UK

William Allsop
Coastal Structures, HR Wallingford, Wallingford & University Southampton, UK

Dave Robinson
Coastal Structures, HR Wallingford, Wallingford, UK

Ingrid Chavet
University College London, Civil & Environmental Engineering, London, UK

Pierre-Henri Bazin
Coastal Structures, HR Wallingford, Wallingford, UK
Ecole Central Lyon, France

Historically, some of the most devastating coastal floods have been those caused by tsunami waves generated by earthquakes, underwater landslides, volcanic eruptions or major debris slides. Tsunami waves travel across oceans with quite small vertical displacements, but shoal up dramatically in coastal and nearshore depths. Generation and transformation of tsunami waves can be simulated by a range of numerical models from source to nearshore, see example results from studies by HRW, POL and BGS on: http://www.defra.gov.uk/environ/fcd/studies/tsunami/default.htm.

Critical gaps in knowledge remaining are in the propagation of tsunami waves in the nearshore region, across the shoreline, and inland. These flow processes cannot easily be simplified, and are indeed made more complex by interactions with beaches, sediment, coastal defences, and then around buildings. These processes can however be simulated in hydraulic models, but correct generation of the tsunami wave is essential, including in some instances the characteristic preceding draw-down wave. Conventional paddle generators simply do not have the stroke to reproduce the entire wavelength, which can be up to 10 km at prototype scale.

This weakness is being addressed within the EPICENTRE research initiative (EPSRC grant EP/F012179/1) through collaboration between University College London (UCL) and HR Wallingford (HRW). HRW are constructing the first Tsunami Generator that will be capable of generating a complete Tsunami wave within a hydrodynamic model by adapting the principles of HRW's pneumatic tide generators, to generate multiple waves (viz. the 3–4 peaks in the Boxing Day Tsunami), and ensure realistic wavelengths. The Tsunami Generator is being mounted within a wave flume equipped to measure coastal processes, inundation and wave forces. It will generate tsunami waves which have been previously transformed from deeper water (approx −200 m) to shallow water (approx −20 to −50 m) using a suitable numerical model. Bathymetry in the wave flume will further shoal the tsunami waves over a representative coastal slope though the shoreline and inland, covering a suitable inland inundation area. Measurements of tsunami transformations through the nearshore region will test/validate existing numerical models. UCL and HRW researchers will then examine interactions of the tsunami with representative coastal seawalls, and test effects of tsunami waves (particularly retreating and repeated waves) with seawall and beach. These tests will then be expanded to measure wave/flow forces on representative buildings and to quantify scour potential around those buildings.

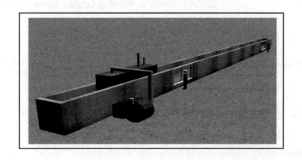

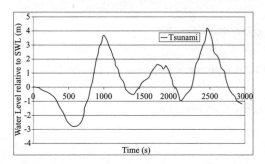

Construction and validation testing of the Tsunami Generator will be completed in summer/autumn 2008. The presentation to this conference will therefore present the first results from this unique flood testing device. As the UCL/HRW researchers expect to complete their tests by spring 2009, the presentation will describe the possibilities for research access to the facility by international teams from summer/autumn 2009.

Keywords: Flood risk assessment, tsunami waves, overtopping, safety analysis

Influence of management and maintenance on erosive impact of wave overtopping on grass covered slopes of dikes; Tests

G.J. Steendam
Infram B.V., Marknesse, The Netherlands

W. de Vries
Infram B.V., Marknesse, The Netherlands
Formerly employed at Water Board Fryslân, The Netherlands

J.W. van der Meer
Van der Meer Consulting, Heerenveen, The Netherlands

A. van Hoven
Deltares, Delft, The Netherlands

G. de Raat
Rijkswaterstaat, Centre for Water Management, Lelystad, The Netherlands

J.Y. Frissel
Alterra, Wageningen, The Netherlands

A huge amount of research is available on the hydraulics of wave overtopping on dikes, levees, seawalls and embankments. In contrast, little research has been performed on the erosive impact of waves overtopping dikes, embankments or levees. Therefore strength of grass covered inner slopes is not well known. With the development of the wave overtopping simulator (Van der Meer et al. 2006, 2007) it is now possible to perform field research on the resistance against erosion of grass covered slopes. First tests with this overtopping device are performed within the ComCoast project (Van der Meer et al. 2007, Akkerman et al. 2007, www.Comcoast.org). For February 2008 further comparative field tests on differently managed and maintained grass covers are scheduled at a Waddensea dike in the Netherlands. These tests will be performed as part of the long term project Overtopping and Strength of Grass Covers within the program Strength of and Loads on Water Defences from the Dutch Ministry of Transport, Public works an Water Management.

At the test location four different test sites are available. All have been treated differently per site for a period of over 15 years. The differences in treatment due to management and maintenance involve whether or not the field is grazed by sheep, whether the field is hayed and differences in amount of fertilizer applied. This has resulted in four different grass covers. The effect of this management and maintenance on the strength against overtopping will be investigated in this project and described in the paper.

All test sites will be tested with the overtopping simulator. Overtopping discharges of 0.1 to 70 or 80 l/s per linear meter dike will be washed down the inner slope of the dike. These discharges will be released in a manner that the real distribution of individual overtopping waves is realised. The condition of the grass cover will be monitored intensively. All relevant strength parameters, such as soil characteristics, layer thickness, root systems of the grass have been and will be analyzed. Also the hydraulic parameters, such as velocity, infiltration and layer thickness of the overtopping water will be measured and analyzed.

Figure 1. Test location.

Figure 2. Enlarged overtopping simulator (22 m³).

The set up of the tests and the results of the prototype tests will be described in the paper. The results together with further research within the program will lead to a Technical Report and new rules for design and safety assessments against wave overtopping.

Keywords: Prototype tests, Grass cover, Dikes, Overtopping, Overtopping Simulator, Erosion resistance

REFERENCES

Akkerman, G.J., P. Bernardini, J.W. van der Meer, H. Verheij and A. van Hoven, 2007. Field tests on sea defences subject to wave overtopping. Proc. Coastal Structures, Venice.
Van der Meer, J.W., W. Snijders and E. Regeling, 2006. The wave overtopping simulator. ASCE, proc. ICCE, San Diego.
Van der Meer, J.W., P. Bernardini, G.J. Steendam, G.J. Akkerman and G.J.C.M. Hoffmans, 2007. The wave overtopping simulator in action. Proc. Coastal Structures, Venice.

Sea wall or sea front? Looking at engineering for Flood and Coastal Erosion Risk Management through different eyes

J. Simm
University of Nottingham, Nottingham, UK
HR Wallingford, Wallingford, UK

As part of the development of the *Performance-based Asset Management System (PAMS)*, the visual inspection of flood defence infrastructure assets is being improved in order to relate the assessment of their current condition to their likely performance. This has resulted in an improved *Condition Indexing* process involving the assessment of the condition of a set of asset *performance features* (Simm et al, 2006).

This work is now being extended to reflect the social (amenity, visual, etc) performance of the assets (Simm & Samuels, 2006). The significance of an asset will depend on the worldview of the person or group (e.g. community 'insider' vs 'outsider') and may be found at different scales and expressed using different frameworks (e.g. landscape character assessments and architectural perspectives). Schama (1996) and Latour (2005) suggest that assets can encapsulate or gain greater significance at particular moments of history—such as times of innovation, change, and breakdown. The principles of strength, utility and beauty first proposed by Vitruvius suggest that flood and coastal assets offer direct and practical amenity value in their multi-functionality. However, they also suggest that the assets act as symbols of things less tangible, offering by interrelation part of a 'realm of significance' for place-based communities, despite the inevitable differences of attitude and approach within such communities. The recent enthusiasm for installation of coastal sculptures in the United Kingdom at places like Newbiggin, Crosby and the Isle of Wight seems to suggest a growing yearning for this broader view to be recognised.

Work is now underway to draw together and test this thinking through a process of conducting interviews with both professionals and community members and activists. The paper will present some outcomes of this work and will challenge the conference to think differently about the structures we spend so much time viewing from an engineering perspective.

Keywords: social performance, asset management, condition assessment, community

Flood Risk Management: Research and Practice – Samuels et al. (eds)
© 2009 Taylor & Francis Group, London, ISBN 978-0-415-48507-4

The new turner contemporary gallery – an example of an urban coastal flood risk assessment

H. Udale-Clarke, W. Allsop & P. Hawkes
HR Wallingford, Wallingford, Oxfordshire, UK

P. Round
Kent County Council, Maidstone, Kent, UK

The new Turner Contemporary Gallery at Margate forms part of Kent County Council's plans to regenerate Margate and the wider district of Thanet. The project has support of many stakeholders, including Thanet District Council, Arts Council England, South East England Development Agency and the European Union. The Turner Contemporary Gallery will commemorate the renowned British artist JMW Turner (1775–1851), who had strong links with the town, and will provide a venue for new work by contemporary artists. As well as several gallery spaces, there will be a large, multi-function space with a terrace overlooking the sea, studio space, café, offices, storage and a workshop. The gallery is to be completed by 2010 and plans were submitted for planning in November 2007.

The gallery is to be built on a seafront subject to storms. Under national planning policy, flood risks associated with new developments in England must be assessed, considering risks to the development itself and changes in flood risks elsewhere. Having previously completed engineering studies on wave overtopping and consequential effects, HR Wallingford was appointed by Kent County Council to undertake a site-specific flood risk assessment for the Gallery to accompany the planning application.

The proposed Gallery is set back from a seawall with open views of the North Sea. The site is partially within the 1:200-year or 0.5% annual probability flood extent defined by the Environment Agency. The building location is however slightly elevated and the Gallery is on a plinth, so is high enough not to be flooded by still water level alone. The main potential flood hazard comes from waves in combination with high sea levels, causing overtopping. Although overtopping water eventually drains back to sea, the movement and accumulation of water during heavy overtopping present a frequent (i.e. return period of 1 in 1-year) flood risk to people, vehicles and operations on the edges of the site. This is observed on the site at present, and is reproduced in desk study calculations of overtopping rates and in physical model measurements of overtopping and water movement.

This paper describes the flood risk assessment undertaken for this urban coastal site and additional studies undertaken prior to the assessment, principally on wave and overtopping (including joint probability analysis and overtopping consequences) and three-dimensional physical modelling of the proposed building and seawall (including measurement of wave pressures and overtopping volumes). Specific reference will be made to HR Wallingford's contributions to Task 2 of FLOODsite, Estimation of extremes, and the new EurOtop Manual http://www.overtopping-manual.com/. The paper will therefore illustrate state-of-the-art scientific developments and methods being applied in the context of producing a formal planning document, for design and assessment of a culturally interesting structure to be built as close to the coastline as possible without actually being in the sea.

Keywords: Flood risk assessment, wave overtopping, wave loading, joint probability analysis

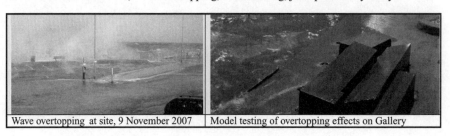

| Wave overtopping at site, 9 November 2007 | Model testing of overtopping effects on Gallery |

EurOtop – overtopping and methods for assessing discharge

T. Pullen & N.W.H. Allsop
Coastal Structures, HR Wallingford, Wallingford, UK

T. Bruce
Division of Engineering, Edinburgh University, Edinburgh, UK

A. Kortenhaus
Leichtweiss Institut für Wasserbau der Technischen Universität Braunschweig, Braunschweig, Germany

H. Schüttrumpf
Lehrstuhl und Institut für Wasserbau und Wasserwirtschaft, RWTH Aachen University, Aachen, Germany

J.W. van der Meer
Van der Meer Consulting B.V., Heerenveen, The Netherlands

1 INTRODUCTION

This paper will describe the new Wave Overtopping Manual (EurOtop, Pullen et al., 2007) developed for EA/Defra, Rijkswaterstaat in the Netherlands and the German Coastal Engineering Research Council (KFKI). The new manual extends and updates the EA's Overtopping Manual (W178) (Besley, 1999), the Netherlands TAW manual (Van der Meer, 2002), and the German Die Küste (EAK, 2002). Considerable research since those publications prompted the production of an updated and extended manual combining European expertise. Research for Defra and the Environment Agency, by HR Wallingford, has provided techniques for predicting the overtopping discharges and consequent flood volumes for a range of seawall types. In the Netherlands and Germany there has been continuous research into overtopping at embankments and dikes, and the European research project CLASH (de Rouck et al. 2005) has expanded understanding of overtopping and scale effects.

2 THE MANUAL

EurOtop incorporates new techniques to predict wave overtopping at seawalls, flood embankments, breakwaters and other shoreline structures facing waves. Supported by web-based programmes for the calculation of overtopping discharge and design details, EurOtop's appendices include case studies and example calculations. EurOtop has an author team of experts in the field of overtopping from several European countries, and has been supervised by a steering committee representing owners and designers from across Europe. EurOtop presents the latest techniques and approved methods for establishing overtopping hazards and flooding for an extensive range of structure types.

3 THE EMPIRICAL METHODS

Empirical wave overtopping methods for coastal dikes and embankment seawalls are discussed, along with Rubble Mound structures. Specifically this covers overtopping for rubble mounds and armoured slopes and uses the techniques described for embankment structures. It includes new coefficients for a much wider range of armour types. Vertical structures extend to vertical, steeply battered and compound vertical structures, and considerable new research on these is discussed. Each of the empirical methods describes how deterministic and probabilistic overtopping assessments can be made, and the degree of any uncertainty that should be considered. Methods are described that can be used to adjust predictions to allow for any scaling or wind effects.

4 CALCULATION TOOL

Accessible from the website, there is an online Calculation Tool that guides the user through a series of steps to establish empirical overtopping predictions for: embankments and dikes; rubble mound structures; and vertical structures. By clicking on graphical representations of structure types and structural features, and by adding the dimensions of the geometric and hydraulic parameters, a range of outputs will be calculated and displayed online. Mean overtopping discharges, overtopping volumes and the percentage of overtopping waves will be displayed. Where appropriate, the calculation of flow velocities and depths at the crest of waves overtopping sloping structures will be given.

REFERENCES

Besley, P. 1999 Overtopping of seawalls—design and assessment manual, *R & D Technical Report W 178, Environment Agency, Bristol.*

de Rouck, J., Geeraerts, J., Troch, P., Kortenhaus, A., Pullen, T. & Franco, L. 2005 New results on scale effects for wave overtopping at coastal structures, *Proc. Coastlines, Structures & Breakwaters 2005, pp29–43, ICE London, Thomas Telford.*

EAK: 2002 Empfehlungen des Arbeitsausschusses Küstenschutzwerke. *Die Küste. H. 65*

Pullen, T., Allsop, N.W.H., Bruce, T. Kortenhaus, A. Schüttrumpf, H & van der Meer, J. 2007 EurOtop—Wave overtopping of sea defences and related structures: Assessment manual. *www.overtopping-manual.com*

Van der Meer, J: 2002 Wave run-up and wave overtopping at dikes. Technical Advisory Committee for Flood Defence in the Netherlands (TAW). Delft. 2002 Keywords

Keywords: Wave overtopping, seawall, coastal flooding, flood risk assessment

Reliable prediction of wave overtopping volumes using Bayesian neural networks

G.B. Kingston, D.I. Robinson, B.P. Gouldby & T. Pullen
HR Wallingford, Wallingford, UK

Climate change and rising sea levels are increasing the risks to coastal flood areas. In order to appropriately protect against such risks, reliable predictions of wave overtopping volumes at coastal structures are required. However, due to the complexity of the physical processes involved and the highly nonlinear dependence between the wave-structure characteristics, this is a nontrivial problem. The accuracy of overtopping predictions is often further hindered by the difficulty in obtaining accurate measurements of the large number of variables. As such, it is of utmost importance to characterise the uncertainty associated with overtopping predictions, as suppressing this information can create a false sense of security in the predictions generated. As a consequence, coastal structures may be inappropriately designed and maintained; in turn, resulting in dangerous conditions in cases of storm surges, wave attack and flooding.

While the importance of providing estimates of prediction uncertainty is increasingly being recognised in the analysis of flood defences and more widely flood risk analysis, it is essential that the derivation of these estimates is as robust and reliable as possible. In this paper, a combined bootstrapping and genetic algorithm technique for obtaining accurate estimates of prediction uncertainty is presented. For the complex, nonlinear models necessary for modelling wave overtopping volumes, classical methods for estimating prediction limits, based on a linear approximation of the nonlinear model function, are unsuitable. Bootstrapping methods provide a nonparametric and relatively straightforward alternative to such methods for uncertainty estimation. However, a limitation of these methods is that they are prone to overestimating prediction uncertainty if care is not taken to minimise model calibration errors. Bootstrapping methods, which involve re-calibrating a model many times to obtain a range of model parameter values that capture uncertainty, have traditionally been used in combination with gradient-based model calibration methods. However, gradient-based techniques can easily become trapped in local minima in the error surface, rather than converging on the global solution. Therefore, rather than reflecting the true uncertainty in the model parameters, and consequently, in the model predictions, the range of parameter values obtained by bootstrapping also captures errors that arise from inappropriate calibration of the model. Therefore, in this paper, a genetic algorithm (GA) is used to calibrate the wave overtopping model, in order to better estimate prediction uncertainty using bootstrapping. GAs are less susceptible to becoming trapped in local minima than gradient-based methods and can, therefore, overcome the problem of uncertainty overestimation. The proposed methods are applied to several example coastal structures and results are compared to those obtained when gradient-based calibration is applied.

Keywords: Wave overtopping, prediction, uncertainty estimation, bootstrapping, genetic algorithms

Calculation of fragility curves for flood defence assets

J.W. van der Meer
Van der Meer Consulting, Heerenveen, The Netherlands

W.L.A. ter Horst
Delft University of Technology, Section of Hydraulic Engineering, Delft, The Netherlands
INFRAM BV, Marknesse, The Netherlands

E.H. van Velzen
Rijkswaterstaat, Centre for Water Management, Lelystad, The Netherlands

In the Netherlands the WV21-project (Water Safety in the 21st century) aims to deliver an up-to-date flood management policy tool. One of the aspects discussed within the project is to gain insight in the effectiveness of measures to mitigate flood risk in future. To be able to perform a quick-scan of the effectiveness calculations are made based on schematised fragility curves.

Safety assessments of flood defence assets are increasingly performed with the technique of structural reliability. All parameters, load parameters (hydraulic boundary conditions) and strength parameters (dike characteristics), are taken into account and expressed as stochastic variables. One of these structural reliability methods is to calculate the failure probability (P_f) of a flood defence given a certain water level (h_w). Assembling the failure probabilities for several water levels constructs a fragility curve (Figure 1).

The use of fragility curves for flood defence assets makes it possible to give a fair estimate of what failure mechanism will be dominant for a certain extreme water level, including waves. For example Figure 1 points out that for water levels below overtopping conditions, the piping phenomenon will be dominant. For more severe situations the emphasis lies on wave overtopping (Ter Horst, 2005). Calculations of detailed fragility curves have already been made for an actual sea dike in the Netherlands. In the near future calculations will be performed for dikes along a large lake and for two river dikes.

Fragility curves will be validated using several probabilistic and deterministic software packages, i.e. PC-Overtopping, a Dutch deterministic model to evaluate the amount of wave overtopping given a certain water level and wave conditions, and PC-Ring, a Dutch probabilistic model for estimating the safety of dike sections.

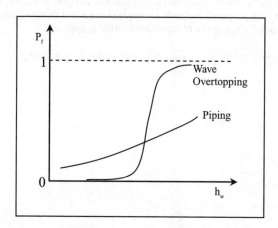

Figure 1. Fragility curves for piping and overtopping.

The paper will give an overall view of the results and validation of the fragility curves for various failure mechanisms i.e. wave overtopping, piping and instability. An attempt will be made to structure fragility curves to a few schematic prototypes, which can be used to speed up calculations on flood probability to a large extent. Furthermore the paper will discuss the typical fields of application, such as estimating the effects of possible flood mitigating measures for policy purposes and the use of fragility curves during flood events using water level predictions (Ter Horst, 2005).

Keywords: Fragility curves, failure probability, flood defence assets

REFERENCE

Ter Horst, W.L.A., 2005, *The safety of dikes during flood waves*, MSc-Thesis, Delft University of Technology, Delft.

Reservoir flood risk in the UK

A.L. Warren
Halcrow Group Ltd, Swindon, UK

The safety of large reservoirs in Britain is promoted by the provisions of the Reservoirs Act 1975. This Act applies to reservoirs that have a capacity in excess of 25,000 m^3. Since the introduction of reservoir safety legislation in 1930, there has been no loss of life arising from dam breach in Britain and there is evidence to suggest that the 1975 Act has further improved the safety of large reservoirs. However, history shows that even small reservoirs can pose a risk to life in the event of dam failure where houses are located close to the reservoir. Recent studies on behalf of the Environment Agency have provided considerable insight to the level of safety provision at smaller, non-statutory reservoirs. Site inspection of over 400 reservoirs throughout England and Wales between 2005 and 2007 was carried out for the purposes of two separate projects. In the first project, approximately 100 non-statutory reservoirs were assessed and ranked in terms of failure risk. In the second project, a general condition score was assigned to over 300 non-statutory reservoirs to reflect the dam safety condition. These projects were carried out in recognition of the Environment Agency's responsibilities under the Civil Contingencies Act 2004 and its role as the enforcement authority for the Reservoirs Act 1975.

The paper aims to:

- Describe and discuss appropriate methods of qualitative risk assessment for large portfolios of reservoirs;
- Consider the risk posed to life and property by non-statutory reservoirs relative to statutory reservoirs;
- Discuss the use of reservoir condition assessments to inform flood risk management;
- Describe the impact of the summer 2007 floods in England on reservoirs; and
- Discuss the possible implications for amendments to reservoir safety legislation in Great Britain.

Keywords: Dam, reservoir, spillway, safety, inspection, maintenance

Flood Risk Management: Research and Practice – Samuels et al. (eds)
© 2009 Taylor & Francis Group, London, ISBN 978-0-415-48507-4

Modelling breach initiation and growth

M.W. Morris & M.A.A.M. Hassan
HR Wallingford Ltd, Wallingford, Oxfordshire, UK

A. Kortenhaus & P. Geisenhainer
Leichtweiß-Institut for Hydraulic Engineering and Water Resources, Technical University Braunschweig, Braunschweig, Germany

P.J. Visser & Y. Zhu
Delft University of Technology, Faculty of Civil Engineering and Geosciences, Delft, The Netherlands

Predicting how a flood defence structure, such as an embankment, behaves under varying load conditions is an essential part of undertaking a flood risk assessment. This understanding directly influences the prediction of rate and volume of any flood water that may cross over or through the flood defence structure and impact on the protected area behind. When the type of risk assessment requires consideration of time varying conditions, the use of a predictive breach model for predicting breach initiation and growth becomes important. Such predictive models are also important for validating simplifying assumptions or equations used in system risk models, where temporal development is not considered in detail, or for any flood inundation models which often require detailed knowledge of the breaching process as input boundary conditions.

Task 6 within the European Integrated Project FLOODsite (www.floodsite.net) contains a programme of work to enhance knowledge and ability to predict breach initiation and growth. This work links closely with other **FLOODsite** initiatives, such as understanding defence structure failure modes and implementing system risk analyses for river, estuary, and coastal areas.

The research that has been undertaken addresses the important process of wave induced breach initiation and builds upon earlier EC research on breach formation under the IMPACT project. The breach formation research also links with an ongoing Canadian/US initiative, facilitated by the Dam Safety Interest Group (DSIG), to review, validate and develop the most promising breach models (worldwide).

This paper will first introduce the wave induced breach initiation research that has been undertaken by the Leichtweiß-Institut of the Technical University Braunschweig (LWI). This research included a review of wave induced breach initiation processes and the development of a model to simulate the processes. Development work was supported by extensive small scale laboratory testing, culminating in the large scale testing of a real embankment section in the GWK flume at Hanover.

The paper then outlines the research and development work undertaken by the Delft University of Technology (TUD) and HR Wallingford Ltd. (HRW) on the development of predictive breach models. TUD developed

Breaching failure of a sea / estuary dike

a new breach model for predicting breach growth through cohesive embankments, building on the earlier work of Visser and using the IMPACT project data sets. HRW developed a second generation version of the HR BREACH model, again including detailed analysis of the IMPACT project data, but also through close collaboration with the DSIG breach modelling project. The paper presents key findings from the research, demonstrates how the models may be applied and subsequently the route through which this may be used within industry to ensure that the latest and most appropriate techniques are used within flood risk analyses.

Keywords: Breach, breach initiation, breach growth, breaching, breach modelling, wave induced breach

A probabilistic failure model for large embankment dams

N.P. Huber, J. Köngeter & H. Schüttrumpf
Institute of Hydraulic Engineering and Water Resources Management, RWTH Aachen University, Aachen, Germany

Planning, construction, operation and maintenance of Germany's large dams are well regulated by law and technical standards. Available dam accident and failure statistics show that the historic performance of german large dams is very high. Nevertheless, due to the high population density in Germany, dam failures pose high risks to society. Meanwhile, risk assessment methods for large dams have been developed and introduced into practical application in many countries worldwide. Thus, dam safety decisions are widely based on risk. The german technical standard for large dams has introduced the demand for a consideration of risk in 2004. Following this development, for german large dams a risk assessment methodology has been developed at the Institute of Hydraulic Engineering and Water Resources Management, RWTH Aachen University throughout recent years. It has found application in different research projects not only on dam safety but also on questions of risk of river dikes and mobile flood protection walls. The focus in the proposed paper will be layed on a quantitative, model-based approach towards the probability of embankment dam failure.

Within the extension of the quantitatively based methodology for risk analysis a special focus was set towards the development of computational tools for assessing probabilities of failures of large dams. The comprehensive probabilistic failure model PrEDaF (**Pr**obability of **E**mbankment **Da**m **F**ailure) was recently developed[1] and allows for the quantification of probabilities with respect to a multitude of complex failure modes of embankment dams. The model covers internal hazards as internal erosion, geostatics and failure of operational equipment, external hazards characterized by extreme floods, earthquakes and landslides and also the effect of insufficient dam monitoring activities. Figure 1 depicts in principle the hazards considered. From these, complex and branching fault trees are derived and mathematically backed by up to date limit state functions as well as numerical modelling approaches, e.g. for reservoir filling and discharge operation in case of extreme hydrological events. Herewith, the embankment dam failure probability can be approximated via detailed probabilistic mathematical-physical modelling of the processes in the catchment area, in the reservoir as well as in a dam body and its subsoil.

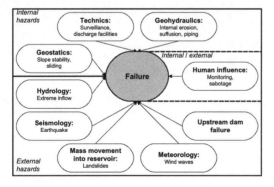

Figure 1. Hazards considered in the probabilistic dam failure model PrEDaF.

---

[1] Huber, N.P. (2007): Probabilistische Modellierung von Versagensprozessen bei Staudämmen. Aachen: Institut für Wasserbau und Wasserwirtschaft, RWTH Aachen. (PhD-thesis, submitted).

Epistemic and aleatory uncertainties relating to dam reliability issues are taken into consideration by applying Monte-Carlo (MC) techniques. This level-III probabilistic approach allows for a future extension of the developed fault trees or replacement of implemented conceptual limit state models in case of the availability of new limit state functions.

During development of the probabilistic model it was especially payed attention to providing a comparable and consistent level of detail for all fault and failure process trees. Due to the representation of probability of dam failure in one numerical value covering all possible hazards this issue was identified as a key demand in probabilistic dam failure modelling.

Focus of the proposed paper is the description of the conceptual framework of the probabilistic model, its high level of detail concerning the different failure modes and the brief presentation of its application in a case study.

Keywords: Dam risk assessment, dam failure, systems analysis, probabilistic modelling, level-III approach

Flood Risk Management: Research and Practice – Samuels et al. (eds)
© 2009 Taylor & Francis Group, London, ISBN 978-0-415-48507-4

Reliability analysis of flood defence structures and systems in Europe

Pieter van Gelder, Foekje Buijs, Wouter ter Horst, Wim Kanning, Cong Mai Van,
Mohammadreza Rajabalinejad, Elisabet de Boer, Sayan Gupta, Reza Shams & Noel van Erp
TU Delft, Delft, The Netherlands

Ben Gouldby, Greer Kingston, Paul Sayers & Martin Wills
HR Wallingford, Wallingford, Oxfordshire, UK

Andreas Kortenhaus & Hans-Jörg Lambrecht
LWI Braunschweig, Braunschweig, Germany

In this paper, the reliability analysis of flood defence systems is outlined. Flood defence structures are used to protect flood prone areas against potential flooding from either rivers, estuaries, or coasts. Probabilistic methods provide a systematic way to assess the flooding probability P_f of engineering structures considering the failure mechanisms of such structures and the uncertainties inherent in the various parameters used for design. For this reason, there is a growing interest in the use of these methods in the design and safety analysis of flood defences and a separate task on this issue within **FLOODsite** (Task 7) has been completed. This paper will first give an introduction to probabilistic analysis, uncertainties, and a reflection on the deterministic versus the probabilistic approach. It will then continue by addressing the tools for a probabilistic system analysis and appropriate calculation methods. Failure probability calculations for an element and a system are also reviewed and a case study of a flood defence system (pilot site of **FLOODsite**) is presented.

A simple example of a flood defence system protecting a flood-prone area is given in Fig. 1. Failure of any of the subsystems (dike, dune sluice, levee) of the system leads to flooding of the floodplain area. All subsystems consist of elements. Sea dikes can for instance be divided in sections. Failure of any of the elements of the subsystem "dike 1" leads to inundation of the floodplain. For all elements of the flood defences, all possible failure modes can be the cause of failure.

The most important failure modes (often captured mathematically through Limit State Equations (LSE) have been reviewed and further elaborated by Task 4 of **FLOODsite**. Within Task 7 the associated LSEs have been developed and coded for use in a reliability tool which is able to calculate the failure probability of the

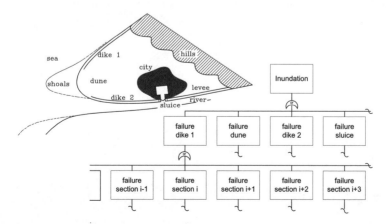

Figure 1. Flood defence system and its elements presented in a fault tree.

top event in the fault tree using a Monte Carlo simulation approach. This tool may be used for any construction type, configuration of the fault tree, and for all probability density functions of load and resistance variables. Load and resistance variables of all structures types have been reviewed and a database of parameters including their uncertainties has been set up and published.

This paper will show the applicability of the reliability tool to several pilot site case studies of **FLOOD*site***. Furthermore, it will address requirements of the European Flood Directive and how the aforementioned results can assist in applying the Directive.

Keywords: Flood defences, reliability analysis, failure probability, systems analysis, Monte Carlo simulation

PCRIVER—software for probability based flood protection

U. Merkel & B. Westrich
Institute for Hydraulic Engineering, Stuttgart, Germany

A. Moellmann
Institute for Geotechnical Engineering, Stuttgart, Germany

In most countries conventional design and dimensioning of dikes is based on predefined hydrological and hydraulic values described and specified by national safety guidelines. Such deterministic design suffers statistical uncertainties and does not allow reliability and failure probability of structures. Probabilistic design of flood defence structures provides a significant contribution to structural risk assessment and cost effectiveness.

A probabilistic design concept, which considers the statistical of design parameters, is core of the PC-River project, which is part of the German research activity "Risk Management of Extreme Flood Events" (RIMAX). Originally this concept was established for dikes within the software PC-RING by the Dutch Rijkswaterstaat which is in charge of all Dutch flood protection measures.

The probabilistic method, which is used by mechanical and structural engineers for design since decades, enables more accurate comparison between dike sections and, therefore more reliable identification of weak points.

The transfer of the concept to inland rivers in mountainous areas with a large variety of hydraulic and hydrological particularities requires special adaptation as shown by an example of the River Elbe in Saxony.

The main advantage of the Netherlands is the quite long lead time for incoming floods as a result of the very large catchment area of the river Rhine. Therefore, the forecasting of water levels is quite accurate. Additionally, the large catchment areas level local extreme values over the distance.

Uncertainties in forecasting the water levels along river dikes in smaller catchments and near to mountainous areas increase, due to:

- spatial and temporal variability of meteorological & hydrological conditions
- weaknesses of hydraulic models (e.g.: changing friction parameters throughout the vegetation period or morphological effects, a.o.)
- unexpected failure of protection measures
- unexpected failure of bridges, landslides etc.

For a probabilistic analysis of a single dike section the uncertainty of the water level is of eminent importance. Especially for the meandering Elbe River the spatial distribution of hydraulic parameters such as water level and flow velocities must be provided by a 2D hydrodynamic numerical model. As the possible combination of all these parameters can not be solved analytically, a Monte Carlo Simulation is the best method to simulate superimposing effects.

2D Models and Monte-Carlo-Simulations are both time consuming procedures. Therefore, a resource saving optimization of 2D-HN-Models became necessary. Multiple adaptive algorithms were integrated in one software package which is able to create ready to use 2D-meshes with defined quality and maximum performance.

Compared to traditional design methods, the results of the probabilistic method show dike sections which need reinforcement. Others turn out to be oversized in relation to the mean of all measures. Hence, new dikes can be designed more economically with respect to reliability and risk assessment.

Keywords: Flood Risk Management, Probabilistic Design, Adaptive 2D-HN Model, Monte Carlo Method

Representing fragility of flood and coastal defences: Getting into the detail

J. Simm, B. Gouldby & P. Sayers
HR Wallingford, Wallingford, Oxfordshire, UK

J-J. Flikweert
Royal Haskoning, Peterborough, UK

S. Wersching
Halcrow Group, Peterborough, UK

M. Bramley
Independent Advisor—Defra/EA Flood Risk Science, UK

Performance-based management of flood and coastal defence assets requires a good representation of defence fragility if investment is to be prioritised on a risk basis. Generalised fragility representations by major asset types can, and have been, used for national and regional scale systems analysis, but more reliable policy and decision making can be achieved if they are more accurate. When making local asset management decisions, accurate site/structure-specific representations become critical.

The paper will describe use of generalised fragility curves in an interim exercise to improve guidance on setting target condition grades. Structures with a sufficiently comprehensive dataset were selected from NFCDD and generalised fragility curves were associated to them according to their structure type and condition grade. The required condition grade for each structure type at a loading equivalent to its declared Standard of Protection was then assessed, assuming conditional failure probabilities of either 1% or 5%. The results to be presented in the paper suggest that current target condition grades are probably conservative, so long as the generalised fragility curves are realistic.

Improving fragility information is being addressed in a number of ways. The principles of reliability analysis are being used in the FLOODsite and FRMRC projects to develop site specific fragility curves, based on a Monte Carlo analysis of reliability for multiple potential failure modes linked by structure-specific fault trees. However, to develop the connection between deterministic and probabilistic approaches (and practitioners), work has also been carried out under the PAMS project to find ways of estimating defence fragility starting from normal UK design practice and Dutch understanding of conditional failure probabilities under design load conditions. The paper will present conclusions from this work, giving some example comparisons between deterministic design and full reliability analysis.

Keywords: fragility, reliability analysis, performance, asset management, asset condition

Flood Risk Management: Research and Practice – Samuels et al. (eds)
© 2009 Taylor & Francis Group, London, ISBN 978-0-415-48507-4

Application of 3D serious games in levee inspection education

Micheline Hounjet, Jos Maccabiani & Rens van den Bergh
Deltares, Delft, The Netherlands

Casper Harteveld
TUDelft, Delft, The Netherlands

A large part of the Netherlands is protected from water by levees. Every five years these levees are checked to find out whether they are still high and strong enough. However, during a high-water situation, there is no guarantee that levees will actually withstand the forces of high water and wind. Therefore, during high water calamities, levee patrollers check the levees for signals that indicate failure mechanisms.

Fortunately, these high water calamities do not occur very often, but when a levee is breached in such a situation, the consequential losses are enormous. Because the occurrence is low, levee patrollers do not have a chance to get fully experienced in levee inspection. It is important that they are able to recognise failure mechanisms in an early stage and that they can communicate about the signal accurately and univocally.

To solve the lack of training opportunities, Levee Patroller was developed by Deltares and the Dutch water boards. Levee Patroller is a 3D serious game, designed for levee inspection training and used by water boards throughout the country to train their own levee inspectors.

In Levee Patroller the player can inspect 3 different 3D regions, based on Dutch polders (figure 1a). The object is to find failure mechanisms and report them accurately. For each region, characteristics of the exercise can be changed: weather type, responsibility level, types and amount of failure mechanisms. The failure mechanisms are based on actual occurrences in the past decades.

When the inspection and communication is correct, measures are taken to prevent an actual breach when the situation becomes critical. When the inspection is incorrect, the levee will actually breach (figure 1b). When the game is finished, the player gets a score, which can be improved when the training is done more often and the player is more skilled in recognition and communication.

At the moment, further developments for Levee Patroller are executed: a special region for levee failures during drought, (storm)sounds to increase the involvement of the player, a device to monitor the education of a player.

Furthermore, there is a development in which the player communicates with a real person instead of the computer. This other person is also a player, but has a different role: the decisionmaker of the water board. He needs to decide whether measures have to be taken to prevent levee failures, based on information given by the levee inspector player. This addition would add more realism to the training and will train an entire link in calamity communication.

Keywords: Levee inspection, training, 3D simulation, serious game

Figure 1a & b.   An inspection region in Levee Patroller and a levee breach.

Strategic appraisal of flood risk management options over extended timescales: Combining scenario analysis with optimization

J.W. Hall, T.R. Phillips, R.J. Dawson, S.L. Barr & A.C. Ford
Tyndall Centre for Climate Change Research, School of Civil Engineering and Geosciences, Newcastle University, Newcastle-upon-Tyne, UK

M. Batty
Bartlett Faculty of the Built Environment, University College London, London, UK

A. Dagoumas
Department of Land Economy, University of Cambridge, Cambridge, UK

P.B. Sayers
HR Wallingford, Wallingford, Oxfordshire, UK

The maintenance and replacement of flood defence systems consumes a large proportion of the expenditure that is committed to flood risk management. Given the significance of this expenditure, risk-based appraisal of the costs and benefits (in terms of risk reduction) of major capital works is now customary in many countries. However, significant idealisations are usually involved in risk-based options appraisal. In particular, the number of options under consideration is restricted to a manageably small set.

In fact, if maintenance, upgrade and replacement are all taken into account, the number of options for management of a flood defence system of moderate complexity is potentially huge. It we consider a system with n defence sections, each of which may be subject to m alternative interventions (e.g. "do nothing", "routine maintenance", "upgrade", "replace") on up to q occasions over the appraisal period, then the total number of options is m^{nq}. In this paper we demonstrate the use of genetic algorithms for solving this risk-based optimisation problem. The approach is first verified in a simplified problem for which the optimal solution can be determined by direct search and is then applied to a realistic flood defence system in the Thames Estuary.

The options appraisal is based on a risk analysis of the flood defence system that incorporates:

- Joint probability analysis of hydraulic loading conditions in the estuary.
- Reliability analysis of the flood defence system, which is a series system of independent defence sections, each of which is characterised by a fragility curve.
- Deterioration of the reliability of the flood defence system, at an uncertain rate.
- Rapid inundation modelling for the defended area, to simulate the effects of overtopping or breaching.
- Damage calculations, based upon maps of the location of properties and standard depth-damage functions.

A cost model has been established that calculates unit costs of maintenance and replacement (which has the effect of reducing the conditional probability of flood defence failure), as well as fixed mobilisation costs. Inclusion of mobilisation costs in the optimisation problem tends to disadvantage options that might involve frequent work on a given defence section and favours options that involve working on a number of neighbouring defence sections at the same time.

The optimisation problem is based upon a Net Present Value criterion, evaluated over the appraisal period. Given that the calculation is based upon a calculation of risk, rather than simply upon probability of failure, intervention in the flood defence system will tend to be targeted at areas with high consequences in the event of flooding, as well as at sections of the flood defence system with a high probability of failure.

The optimisation problem is in the first instance set up as an unconstrained problem, but in practice this is unrealistic, as the resources available for maintenance and replacement of flood defences will tend to be constrained each year. A number of alternative approaches have therefore been tested:

1. A strict constraint on annual spend.
2. A constraint on total spend over a number of years and allow some inter-annual variation.
3. Do not apply any constraint, but in the objective function of the optimisation penalise options with uneven annual spend.

Keywords: Risk-based decision making, Flood defences, Reliability, Optimisation, Maintenance

Embedding new science into practice – lessons from the development and application of a Performance-based asset management system

C. Mitchell, O. Tarrant & D. Denness
Environment Agency, UK

P. Sayers & J. Simm
HR Wallingford, Wallingford, Oxfordshire, UK

M. Bramley
Independent Advisor to Defra/Environment Agency F&CERM R&D Programme, UK

The UK Environment Agency is developing an improved performance and risk based approach to asset management associated with flood and coastal defences. This utilises a number of advances made within the research community over the past five-ten years, in particular the so-called PAMS methods—Performance-based Asset Management System. Delivering innovation into practice however requires significant effort, and this effort is often under-estimated.

This paper explores the barriers and facilitators of moving innovative and potentially beneficial science into good practice. This is done with reference to specific examples drawn from the development of the PAMS including issues associated with achieving buy-in from multiple users. For example take-up can be undermined by researchers through over-selling of the utility and readiness of the science as well as by potential users through resistance to change and perceived loss of commercial advantage. An important means of overcoming these hurdles is the process of piloting and independent verification of the methods. This paper explores how close working with specific end-users within the Thames Estuary Strategy team as well as national and local Environment Agency asset management staff has helped build trust in the science and demonstrate its benefits.

Access to the new tools, user skill and training, and crucially the on-going support for this, also plays an important part. Through the UK Conveyance Estimation System (CES) the research community has provided a step change in the ability of managers to explore the impact of changes to management practice on channel performance. Delivering the CES tool into practice has highlighted a number of specific difficulties, that if not addressed, would prevent effective take-up. For example, liability and ownership; support and training; associated policy guidance.

Where new tools challenge existing paradigms, take up can be slow. There is an onus on researchers to engage with, and demonstrate utility to, industry. For example PAMS includes a move towards reliability-based assessment of asset performance under load. Concepts of fragility curves, where performance is considered probabilistically across a wide spectrum of loading, challenges a more traditional deterministic consideration of design loads. This paper highlights how close working with engineering practice through examples has helped explore these issues and promote take-up; it also highlights how difficult this road can sometimes be.

Other issues explored in the paper include the data needs of new tools and perceived complexities of new methods, as well as Information Technology (IT), Intellectual Property Rights (IPR) and liability.

Through the paper examples are used to illustrate the points made. It concludes with a series of useful recommendations to maximise the success of the uptake process, and avoid even the best science from being left on the shelf.

Keywords: Asset management, Performance, Piloting, End user

Study of flood embankment behaviour induced by air entrapment

D. Lesniewska
Polish Academy of Sciences, Institute of Hydro-Engineering, Gdansk, Poland
Koszalin University of Technology, Koszalin, Poland

H. Zaradny
Polish Academy of Sciences, Institute of Hydro-Engineering, Gdansk, Poland

P. Bogacz & J. Kaczmarek
University of Warmia and Mazury, Olsztyn, Poland

Unsaturated soil is a multi-phase system consisting of grains and voids containing air, water vapour and liquid water. The role of air pressures in typical geotechnical conditions is often neglected, as the characteristic time of typical structure loading is usually much longer than the dry consolidation time. However, there exist situations when much shorter loading times appear, like sudden flooding, when the coupling between air, water and soil grains has to be taken into account. Most of the study reported here was performed as a part of FLOODsite IP (Task 4, Report Number T04-07-10). It was focused on performing a series of infiltration tests on the models of flood embankment. The models had inclination 1:2 and were 0.8 m high and 1.60 m wide. They were built by pouring sand from hoppers. Pressure sensors were installed on the front wall of the test box to measure soil water and air pressures. The tests were recorded using digital cameras to deliver data for PIV analysis. It was

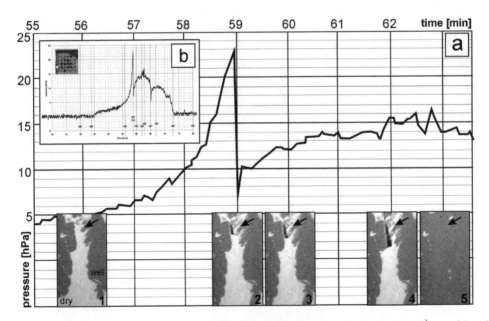

Figure 1. Relation between soil air pressures and cracks forming within the flood embankment body, a) part of the soil air pressure graph, related to the test period when cracks emerging was observed. The pressure measurements were made by the sensor located close to the area of air closure (photographs 1–5, taken between 56 and 63 minute of the test). Location of the crack is marked by an arrow, b) complete air pressure measurements made by the same sensor.

confirmed during the tests that overflow or wave overtopping, if sufficiently intensive, could leave macroscopic areas of unsaturated soil, where a part of soil air was entrapped. Measurements held during the tests proved that the entrapped air pressure was higher than the atmospheric one. The measured difference was about 10 hPa in average and about 25 hPa at peak (see Fig.1). It was enough to induce internal cracks and cause their opening (Fig.1). In some cases rapid air escape to the atmosphere was observed. Escaping air usually caused local surface damage. Cracks appeared within the entrapped air areas, always near a border between dry and saturated soil. Clear embankment failure mechanism (micro-slides progressing up along its slope) was observed during the tests. It was stated later, with a help of PIV image analysis (displacement and strains), that in fact only a part of global failure mechanism could be visible with a naked eye. It was also found that the maximum displacements of the embankment concentrated just within the entrapped air pockets.

Keywords: Flood embankment, model tests, air entrapment, cracks, failure

Assessment of flood retention in polders using an interlinked one-two-dimensional hydraulic model

M. Kufeld, H. Schüttrumpf & D. Bachmann
Institute of Hydraulic Engineering and Water Resources Management, RWTH Aachen University, Aachen, Germany

An effective measure to reduce the impacts of floods is the retention of the flood waters to limit the maximum discharge. In the upper reaches of a river, dams may provide the required storage volume. In the lowland region the flood waters can be held back in diked areas next to the river: Polders. These polders are flooded as the flood wave propagates downstream.

In a study conducted for the German RIMAX ("Risk management and assessment of extreme floods") joint project *Flood Control Management for the River Unstrut*, the functionality of an existing flood control system was evaluated and solutions sought to enhance the capacity to reduce the impact of floods.

In a probabilistic approach different system states and hydrological loads are investigated. For each discharge of a certain average recurrence interval, a set of different hydrograph shapes is considered. The different hydrological loads are boundary conditions for the hydraulic model, which then is adapted to represent the different system states (i.e. present state, enhancing of existing polders, additional new polders, controlled and uncontrolled flooding etc.). A multitude of scenarios have to be calculated, which requires the model to perform as fast as possible, while maintaining the accuracy needed to allow sound conclusions.

To meet these requirements a model code was developed that couples a one-dimensional hydrodynamic and a two-dimensional storage cell model. The flow in the main channel is represented using a diffusive wave approximation. Overbank flow is modelled by storage cells with representation of momentum transfer through Manning's equation. Both modules are bidirectional linked by empirical hydraulic relationships, which describe a direct relationship between discharge and flow depth, i.e. the weir equation for flow over a dike.

Storage cell models perform fast, but as they are based on the solution of a set of stiff ordinary differential equations, explicit numerical schemes require a very small time step to guarantee stability, thereby losing some of this computational advantage. Here an implicit time integration scheme was used to allow an error based adaptive control of the time step length, so that the optimum of computational time and accuracy can be achieved.

From the results obtained in this study conclusions are being drawn regarding the model performance and the effectiveness of flood retention in polders.

Keywords: Flood inundation modelling, flood defence infrastructure, lowland river floods

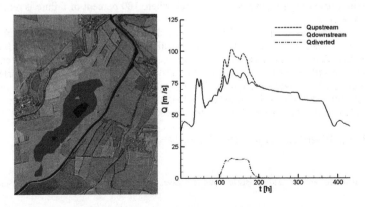

Figure 1. Aerial view of simulated polder flooding and resulting hydrographs.

Fragility curve calculation for technical flood protection measures by the Monte Carlo analysis

D. Bachmann, N.P. Huber & H. Schüttrumpf
Institute of Hydraulic Engineering and Water Resources Management, RWTH Aachen University, Aachen, Germany

In the framework of the multidisciplinary research project "REISE", supported by the German BMBF-National research programme "Risk Management of Extreme Flood Events" (RIMAX), a multi criteria decision support system is developed in order to assess different protection measures and, thus, to support the preliminary design of flood risk management plans on the rivershed scale. The main criterion for the assessment of flood protection measures is their effect on the flood risk. It is defined as the probability of flooding multiplied with the potential damages.

For the determination of the flood risk on the rivershed scale an integrated approach is applied. The whole rivershed and its flood protection line are considered as a system with possible system states[1]. The flood protection line is divided into sections. Possible system states range from no flooding in the area to a failure and consequential a flooding of all sections. Theoretically all system states can be examined by an event tree analysis: the probability of the occurrence of flood, the failure or non-failure probability of the flood protection sections and the resultant consequences by flooding of the sections are logically combined to the flood risk of the whole rivershed. The major challenge of this approach is the very time-consuming computation of the flood risk. This problem gets even more relevant if the flood risk should be used as a criterion to evaluate the effectiveness of different flood protection measures. An acceptable computation time is required.

The idea is to pre-evaluate the variables of flood risk for a rivershed and save them in a database. The probability of the occurrence of flood is stored in a hazard curve, the performance of the flood protection line and with it the probability of failure or non-failure for a section is expressed by the fragility curves. The flooded area and the resulting consequences are also saved in a database for discrete cells. By virtually implementing and verifying a flood protection measure just the affected curves and data have to be re-evaluated and newly combined to the flood risk by an event tree analysis.

The presented paper focuses on the fragility curve, one part of the variables of flood risk. It shows the probability of failure in dependency of the stress applied to the protection measure. Here, the stress is defined as the water level in the section of the flood protection line.

The fragility curve can be quantified by help of expert judgment or by a direct calculation of the failure probabilities. A direct calculation of the fragility curve is proposed: a Monte-Carlo analysis is applied for discrete water levels in the section, which results in a failure probability for a section of the protection line. By reapplication of the Monte-Carlo analysis until the water level, where 100 percent of failure is reached, a discrete fragility curve is obtained.

In the proposed paper the fragility curve calculation for a dike as an example of technical flood protection structures are presented. Besides the demand of a fast calculation of this curves, technical flood protection measures as a dike strengthening, e.g. a dike covering by grass, must have an effect to their calculation. The fragility curve is the main path, how their effect to the flood risk can be evaluated. Thus, appropriate geostatical, geohydraulic and hydraulic limit state functions are derived and their combination in a fault tree are shown. The effect of dike strengthening methods and their influence to the determination of the failure mechanism are discussed. First results, how different types of flood protection measures effect the fragility curve will be shown.

Keywords: Flood Risk, Rivershed, Fragility Curve, Flood Protection, Limit State Functions

---

[1] Hall et al., 2005: A methodology for national-scale flood risk assessment; Stoch Environ Res Risk Assess 19; 388–402; Springer.

Application of GMS system in the Czech Republic – practical use of IMPACT, FLOODSite and GEMSTONE projects outcomes

Zuzana Boukalová
VODNÍ ZDROJE, a.s., Prague, Czech Republic

Vojtěch Beneš
G IMPULS Praha spol. s r.o, Prague, Czech Republic

GMS system (Geophysical Monitoring System) has been designed within IMPACT and FLOODSite projects as a new instrument for maintenance of and checks on safety of flood control dikes. Checks on safety of dikes performed so far have particularly been based on visual inspection in situ or on analysis of airborne and satellite photographs. In this way, only larger defects that are manifested on dike surface (dike shape deformations, permanent seepage) can be discovered.

The main objective of the geophysical measurements within GMS is to provide "hidden" information on inner structure (defects) of dikes and materials used for their construction. Thanks to this, we can design the optimal process of dike reconstructions and remediations within the entire river basin areas. Repeated measurements under different hydrological conditions help to identify seepage paths through the dikes or their underlying layers. The measured data are filed in the database, which in an appropriate form through the Internet is made available to the responsible persons in River Basin Organizations. For these applications, special programmes GMS_Data (interpretation programme) and GMS_view (database viewer) have been prepared.

At present, GMS system in the Czech Republic is being introduced in 3 of 5 river basin areas. Since 2004 (starting year of GMS system introduction), approx. 200 km of flood control dikes in the Morava River Basin Area, approx. 80 km of dikes in the Odra River Basin Area and approx. 10 km of dikes in the Labe River Basin Area have been measured. These are the largest risk posing segments which regulate watercourses in the vicinity of bigger towns. The database has been growing, at present it includes approx. 15% of the total volume of the dikes in the Czech Republic.

The presentation demonstrates the ways of GMS application including software as well as the results gained so far in the process of system introduction in practice in the main River Basin Areas in the Czech Republic.

Keywords: Flood risk management, flood control dikes, geophysical measurement, monitoring

Failure modes and mechanisms for flood defence structures

M.W. Morris & W. Allsop
HR Wallingford, Wallingford, UK

F.A. Buijs
Delft University of Technology, Faculty of Civil Engineering and Geosciences, Delft, The Netherlands.

A. Kortenhaus
Technische Universität Braunschweig, Leichtweiß-Institut für Wasserbau, Braunschweig, Germany

N. Doorn
Previously of WL/Delft Hydraulics, Delft, The Netherlands

D. Lesniewska
Institute of Hydroengineering, Polish Academy of Science, Gdanks, Poland

The European Integrated Project **FLOODsite** (www.floodsite.net) will provide better understanding and improved methodologies for flood risk analysis and flood risk management. This includes consideration of flood risk from river-basins, estuaries and coasts, from a range of perspectives and users. **FLOODsite** is subdivided into different Themes, one of which (Theme 1) deals with improved understanding of the underlying physics needed to perform a flood risk analysis. Within Theme 1, Task 4 addresses flood defence failure mechanisms and the development of a definitive document detailing different failure modes for a wide range of defence types and loading.

The understanding and analysis of flood defence failure mechanisms was identified as a critical area of research where improved knowledge on processes and their representation through limit state equations was urgently needed. In order to undertake a flood risk analysis, the performance of different flood defence structures under varying load conditions needs to be represented. The work in Task 4 addresses the review, collation and refinement of failure mode information, as well as research into some specific defence failure processes. The objective was to bring together all available information on failure modes, for a wide range of flood defence structures, to support the development and implementation of system wide models for flood risk assessment.

This paper will describe the key elements of research dealing with failure modes for flood defence structures, leading to the production of the FLOODsite report T04-06-01, *Failure Mechanisms for Flood Defence Assets* (available online from www.floodsite.net). This work comprised a review of structures and failure modes which have occurred in the past, investigation of limit state equations and associated uncertainties for both the models and the input parameters, and some additional defence specific investigations to improve knowledge of failure processes (including embankment and shingle beach failure processes). It is envisaged that this report will be maintained and updated in the future.

The report has been structured so as to allow the user to consider structure type and structure loading, and then refer to a summary description, diagram and limit state representation for appropriate failure processes. The structure type and loading is presented through a matrix table, with structure types including (i) **beaches and foreshores**; (ii) **banks** (beaches, dune and shingle banks, embankments, and revetment systems); (iii) **walls**; and **point structures** (culverts and pipes, sluices, barrages, gates and pumps). A range of loading types have been considered, including static water, waves, and lateral loading. Failure modes have been identified for most structures (see example in photo) with about 80 templates provided.

Typical sliding failure of the turf at inner slope of a sea dike

The 'failure modes' work is discussed in more detail in the paper, including details of additional research undertaken to refine or produce new failure mode descriptions. The results have been used as the basis for improved risk analysis methods and tools, feeding directly into a number of the FLOODsite Pilot Site studies. Results will also help in implementing requirements of the European Flood Directive.

Keywords: Flood risk analysis, flood defence structure, failure modes, breaching, breach development

Non-structural approaches (CRUE project)

Flood risk map perception through experimental graphic semiology

S. Fuchs
Institute of Mountain Risk Engineering, University of Natural Resources and Applied Life Sciences, Vienna, Austria

W. Dorner & K. Spachinger
Department of Civil Engineering, University of Applied Sciences, Deggendorf, Germany

J. Rochman & K. Serrhini
UMR CNRS 6173, University of Tours, Tours, France

The procedure of risk assessment emerged as an appropriate tool to analyse the impacts resulting from natural hazards. However, despite from a considerable amount of approaches and guidelines, only little work has been carried out with respect to the harmonisation of risk mapping, the information necessary, and the required quantification of possible impacts on decision-makers. A particular gap exists with respect to the perception of information provided in risk mapping for different stakeholder groups.

Within the RISKCATCH project, risk maps were created for catchments located in the Alps and the related forelands. Based on the assessment of historical and possible future development of hazard, values at risk and vulnerability, these maps were created on different scales using scenario technique. The information created was evaluated by presenting the maps to several stakeholder groups, above all political decision makers, practitioners and laypersons from European countries, using the method of graphic semiology. As a result, possible improvements in design and content of such information were deduced.

The respective hydrological processes (precipitation, land use, geomorphology) and the rivers (structure, flood plain and flood defence) were analysed on a regional and temporal scale for three catchments in the Austrian Alps and two catchments in the German alpine foreland. Existing data of synthetic floods derived from hydrological models in combination with stream routing was linked with data of real flood and debris flow events to analyse the relations between precipitation and the development of hazards in the river system. Based on the modelling results of defined design events, the associated damage potential was assessed. Multi-temporal data on damage of previous inundations and data on the associated damage potential in the test sites was collected and edited for the use within a GIS-environment. The elements at risk—which were defined as those buildings within the test sites located in endangered area—were analysed with respect to their spatial location and extension using GIS. Intersecting this information, risk maps were generated on different scales in order to analyse possible improvements in information content and design.

To quantify risk perception, the maps were presented to a group of stakeholders from different European countries. The method used was based on the approach of experimental graphic semiology, reversing the traditional communication pattern from transmitter to receiver. Starting from receiver, the maps were presented to the test persons using an ophthalmic device for the record of eye movements during picture reading. The test was accompanied by a specific survey; hence, the cognitive perception of risk maps was evaluated. All maps were presented to the test persons for a relatively short time period to identify the most attractive components of each map. The eye movements were subsequently statistically analysed in order to assess patterns of visual perception for each map and to study the reading behaviour for text elements included in the maps. The visual strategies of each test person were quantified.

The study resulted in guidelines of how to include information on natural hazard risk for different stakeholders, in particular with respect to visual information contained in risk maps. Thus, the results of the study can be used by administrative bodies for communication strategies within the concept of integral risk management, in particular with respect to non-structural flood mitigation.

Keywords: Flood risk management, risk mapping, experimental graphic semiology, European research

Quantifying the benefits of non-structural flood risk management measures

R.J. Dawson, N. Roche, A.C. Ford, S.L. Barr & J.W. Hall
Newcastle University, School of Civil Engineering & Geosciences, Newcastle upon Tyne, UK

J. Werritty, T. Ball & A. Werritty
UNESCO Water Centre, University of Dundee, Dundee, UK

M. Raschke & K. Thürmer
IWSÖ, Weimar, Germany

There are a wide range of flood risk management measures including flood defences, flood warning, land use regulation, insurance *etc*. Whilst significant advances have been made at quantifying the risks associated with flood defence systems, appraisal of the effectiveness of non-structural measures has hitherto been limited to estimating their influence on parameters in established hydrological, inundation or damage models.

This paper introduces research, funded through the CRUE ERA-NET programme, that quantifies the effect of non-structural measures in coupled land use and flood risk simulations on the long term evolution of flood risk. The emphasis is upon looking at changes in risk over extended timescales due to changes in land use and vulnerability which, in turn are influenced by changes in regulation, such as development restrictions, and market instruments, such as insurance.

Our method builds upon land use modelling capabilities developed in work described by Dawson *et al.* (submitted to FloodRisk 2008). This model couples simulations of regional economy, landuse change and hydrodynamics with a systems-based risk analysis methodology to estimate current and future flood risk (Figure 1). This model, initial development of which has been funded by the Tyndall Centre for Climate Change Research, has been extended to provide an improved description of the landuse market and can consequently be extended to explore the effect of market and regulatory instruments on landuse change. This enables non-structural and structural management measures to be appraised using the *common currency* of risk and can be used to test and evaluate the efficiency and the sustainability of different combinations of structural and non-structural

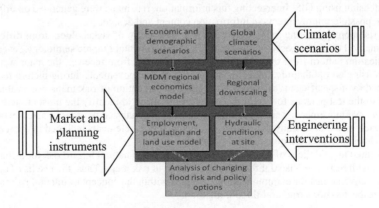

Figure 1. Non-structural flood risk analysis methodology.

measures under different scenarios of economic and climate change. These quantified outputs directly support scheme appraisal, and feed into existing asset management and risk assessment systems.

The paper first provides an introduction to non-structural flood risk management and the role of insurance and regulation in managing flood risk before describing the systems-based approach to quantifying the effectiveness of non-structural flood risk measures (with a focus on insurance mechanisms and landuse planning) and the landuse change model before presenting some results of a demonstration in the Thames Estuary.

Keywords: Non-structural measures, flood risk analysis, land use simulation

Efficiency of non-structural flood mitigation measures: "room for the river" and "retaining water in the landscape"

S. Salazar & F. Francés
Hydraulic and Hydrology Research Group, Technical University of Valencia, Valencia, Spain

J. Komma & G. Blöschl
Institute for Hydraulic and Water Resources Engineering, Vienna University of Technology, Vienna, Austria

T. Blume, T. Francke & A. Bronstert
Institute of Geoecology, Chair for Hydrology and Climatology, University of Potsdam, Potsdam, Germany

Floods are the costliest natural hazard in the world. A review of the losses caused by floods events in the period of the last ten to fifteen years indicate that in Europe economic losses are dramatically increasing, mainly because there has been a marked increase in the number of people and economic assets located in flood hazard zones. Flood risk mitigation can be done reducing vulnerability and/or hazard in order to reduce the total flood impact. The aim of this project was to examine the relative efficiency of non-structural flood mitigation measures. To do this, new scientific frameworks and technical tools integrating multidisciplinary approaches (meteorological, hydrological and hydraulics) were used on three non-structural flood mitigation measures: land use changes (afforestation/deforestation); local retention measures in the landscape through micro ponds, ponds, small dams and a single dam; and flood retention along the main stream by providing inundation area based on the room for the river concept. These scenarios were analysed and tested in three different and, at the same time, complementary case study areas: the semiarid Rambla del Poyo in Spain (380 km^2), the humid/midland Kamp in Austria (1500 km^2), and the alpine/prealpine Iller in Germany (950 km^2). The floods in these catchments cover a wide spectrum of processes to be expected in the European Research Area.

This research summarizes the proposal submitted and approved in the Era-net Crue Funding Initiative last call. The methodological tool consists in the project division into four work packages (WP). In WP1, scenarios have been defined that represent the various mitigation measures as well as "input" scenarios. WP2 focused on the catchment hydrology, and includes parameterisation and verification of distributed hydrological models, performing simulations for the scenarios and a statistical assessment of the simulated events. These simulations are used in WP3, where hydraulic analyses have been performed along the river reaches of interest by running non-steady state models and/or linking up to existing results of hydraulic models. These analyses allow the efficiency assessment of retention measures along the main stream and provide the flood inundation areas. Flood hazard maps have been the main output of this WP. Finally, WP4 assesses the efficiency of the flood mitigation measures in terms of the impact or risk by combining the hazard with the vulnerability of the flood prone areas.

At this moment, the most important results are: For similar total precipitation, the more convective storms produce larger floods. A large number of small dams are less efficient than a single dam with similar capacity in volume and located downstream. Overall the findings indicate that the potential of micro ponds, ponds and small dams as local flood retention measures is moderate to small, depending on the scale of the magnitude. Decision on whether such structures are a suitable retention measure in any one catchment would have to be based on an integrated flood management plan that balances the various options of flood mitigation and management. In comparison, the afforestation scenario represents a better mitigation measure. The efficiency of "room for the river" measures depends on the relative magnitude between the inundation area and the flood volume.

Keywords: Flood mitigation measures, room for the river, retaining water in the landscape

Flood risk reduction by PReserving and restOring river FLOODPLAINs – PRO_FLOODPLAIN

H. Habersack, C. Hauer & B. Schober
University of Natural Resources and Applied Life Sciences, Vienna, Austria

E. Dister, I. Quick & O. Harms
University of Karlsruhe, Rastatt, Germany

M. Wintz & E. Piquette
Université Marc Bloch—Strasbourg, Strasbourg, France

U. Schwarz
Fluvius, Vienna, Austria

The main objectives of this paper are to identify and evaluate the hydromorphological contribution of water retention in preserved and restored floodplain systems across various geomorphological settings (hydromorphology), to demonstrate the contribution of floodplain preservation and restoration to the achievement of the good ecological status (ecology) and the investigation of the social acceptance of such non-structural measures (sociology). The transformation in a general floodplain evaluation matrix is presented on the basis of varying boundary conditions in different study reaches in Austria, Germany and France. The benefits and disadvantages of the floodplain enlargement are compared with each other and considered in comparison with technical measures. In the *hydromorphological* context one and two dimensional hydrodynamic modelling shows the influence of various parameters on the effect of river floodplain inundation. Thereby a multi parameter variation allows a derivation of characteristic relation functions using e.g. retention volume $V_{floodpl}/V_{total}$ [-], $Q_{river}/Q_{floodplain}$ [-], channel slope [-], floodplain width $[m]/Q_{bankfull}/Q_{total}$ [-], depth relation [-], landuse [k_{str}/n—values], sinuosity [-], bed slope, aspect ratio, roughness, inundation volume. Based on this theoretical analysis several technical and non-technical retention areas along the Danube River, March River, Kamp River in Austria and Upper Rhine River in Germany/France have been evaluated for the effects of flood peak reduction, wave translation and retention for flood events with different recurrence intervals ($HQ_5 \longrightarrow HQ_{100}$). As an example one retention area (68.71 km long) at the Danube River leads to a peak reduction of 954 m³s⁻¹ for a 100-years flood, which is about 8% of the maximum discharge.

The *ecological* analysis discusses the ecological value of still intact and cut off floodplains by means of a variety of different morphological, hydraulic/hydrological, floristic and faunistic parameters, using GIS techniques. Using non-technical measures the restoration of floodplains is not restricting water inflow, hydrodynamics, sediment balance and nutrient exchange. The results of the study demonstrate that the reestablishment and regeneration of typical floodplain biotopes occur within a relatively short time frame. This applies to all those biotopes that result from the regular perturbation by erosion and sedimentation as well as an interruption of the succession (e.g. channel shifts). For the creation of these morphological structures a minimum dynamics is necessary, related to flood magnitude, as shown by recent extreme floods. In contrast to short term effects of vegetation reappearance, the development of floodplain forests is a longterm issue. Thus existing, ecologically functioning floodplains are essential and have top priority to be preserved, followed by floodplain restoration, improving mostly simultaneously flood protection. The *sociological* contribution focuses on the analysis of non-structural measures as a way of protecting against flood risk aiming to define the process of social acceptance of these non-structural measures. This paper emphasizes the elements/arguments on which the decision makers support floodplain preservation and restoration and the tolerance level of the population or a concerned group to decrease regulations in favor of non-structural measures. The interdisciplinary outcome of this paper is the floodplain evaluation matrix, integrating technical, ecological and sociological aspects and giving decision makers and practitioners an objective basis for priority judgements and the implementation of the EU Floods Directive.

Keywords: Floodplain preservation and restoration, non structural measures, integrated flood risk management

The use of non structural measures for reducing the flood risk in small urban catchments

E. Pasche & N. Manojlovic
Hamburg University of Technology, Hamburg, Germany

D. Schertzer, J.F. Deroubaix, I. Tchguirinskaia & E. El Tabach
Université Paris-Est, ENPC/CEREVE, Marne-la-Vallée, France

R. Ashley & R. Newman
University of Sheffield, Sheffield, UK

I. Douglas & N. Lawson
University of Manchester, Manchester, UK

S. Garvin
BRE Scotland, East Kilbride, Scotland

Many small urban catchments have no visible surface streams and are affected by overflows from sewers and stormwater drains. Using the possibly controversial definition of non-structural measures (NSM) as *responses to urban flood risk that do not involve fixed or permanent facilities and that usually work by changing behaviour through government regulation, persuasion, and or economic instruments,* parallel assessments of NSM implementation, relevance and performance were made in cities in France, Germany, England and Scotland to help develop practical guidelines and decision making tools. Inventories were made of each city's hydrological and socio-economic situation (flood risk and management plan, regulatory, and cultural conditions). Focus groups and interviews were held with flood victims, and local and national level stakeholders to investigate emergency response, adaptation to flooding, people's risk culture and awareness of soft mitigation measures and NSM. Four NSM categories were used: Awareness of flood risk, Alleviation of flood impacts, Avoidance of the risk if possible, and Assistance in emergencies, based on the approach developed in Scotland.

Until a flood occurs, residents are still often unaware of the risk, even if living near a hot spot for water build-up and sewer overflow. Worse still, many planning authorities seem to be unaware of surface water flood risks, there being too many examples of flooding of properties built in the last 20 years. Awareness raising is thus a key need; further capacity building is required among professionals and public alike. As the UK Pitt Review emphasised ("Learning lessons from the 2007 floods" The Cabinet Office, 2007, Ref: 284668/1207) existing flood risk maps often ignore areas affected by pluvial (surface water) flooding in small urban catchments. More refined, reliable, hydrological models, helped by new measurement networks and databases would define this risk better. Flood warnings systems, including radar-based rain prediction, using modern telecommunications, need strengthening to reach all people at risk. Agency responsibilities are divided in all four countries, and local authorities often lack both financial and human resource capacity to take the lead. However, although there are key differences between countries in the relative responsibilities of local and county councils, national environmental agencies, privatized water and sewerage organizations and government or private insurance schemes, the UK Pitt Review interestingly emphasises that city authorities should take the lead in managing flood risk and surface water in small urban catchments. Greater collaboration and openness both between agencies and between the different professions is required. At present, because the risk of flooding in small urban catchments varies greatly in probability, and is largely driven by rare high intensity summer thunderstorm events, flood mitigation measures are few and reliance insurance to cope with losses is high. Yet possession and costs of insurance are highly variable. Adoption of measures to increase building resilience is low and will vary from one householder to another. The likely impacts of climate change are not considered by most stakeholders.

No regret solutions, with the flexibility both to cope with changed storm patterns in the future and to increase resilience to reduce vulnerability, are required. Appropriate combinations of structural, soft engineering and non structural measures, with designed adaptability and reversibility are needed. Further research, monitoring and modelling, with good use of existing data, will help derive more efficient decision-making tools to assess NSM performance and sound implementation.

Keywords: Extreme and risk assessment, mitigation, urban flooding, non structural measures, planning, insurance, capacity building, co-operation

EWASE—Early Warning Systems Efficiency: Evaluation of flood forecast reliability

K. Schröter & M. Ostrowski
IHWB. TU-Darmstadt, Darmstadt, Germany

M. Gocht
Water & Fincance, Berlin, Germany

B. Kahl & H.-P. Nachtnebel
IWHW, BOKU Wien, Vienna, Austria

C. Corral & D. Sempere-Torres
GRAHI-UPC, Barcelona, Spain

EWASE (Effectiveness and Efficiency of Early Warning Systems for flash-floods) is a R&D project within the ERA-NET CRUE integrated project supported by the European Comission under FP6. Medium sized river basins prone to flash floods are considered. The results presented are obtained from two case studies: the Besòs basin (1020 km^2), Mediterranean climate, near Barcelona, Spain, and the Traisen basin (921 km^2), alpine climate, north east Austria.

Flood alerts provided by early warning systems are an important element of comprehensive flood risk management. Indeed, the potential benefit of anticipating imminent floods is unquestioned. However, the utility of the released alerts largely depends on the available lead time to take preventive measures. Therefore, reliable flood forecasts with sufficient lead time are crucial for the warning system to be effective. Both meteorological and hydrological aspects affect the reliability of flood forecasts. In this regard, the capability to reliably predict floods is determined by the anticipation and provision of rainfall information and the description of the hydrological system state and behaviour by means of a hydrological simulation model.

The objective of this work is to analyse the lead time dependence of flood forecast reliability in view of uncertainty associated to precipitation forecasts and hydrological simulation as an important basis to assess the effectiveness of early warning systems.

The lead time dependence of flood forecasts is analysed by means of a multiple step ahead forecast approach. This includes the generation of forecasted hydrographs for multiple time steps t_i ($i = 1,\ldots n$) of an event using different lead time periods τ ($\tau = +1,\ldots,+p$ [h]). Hydrographs are simulated using observed rainfall until t_i and the rainfall forecast available at t_i until $t_i+\tau$. Combining the discharge values calculated for the different points of time $t_i + \tau$ results in a forecasted hydrograph Q_τ, which is conditional on the forecast lead time τ.

The uncertainty inherent to anticipated rainfall information is quantified in terms of a precipitation ensemble forecast. Hydrological model uncertainty is described by application of different hydrological simulation models. The multi model ensemble consists of two continuous (COSERO and WBrM) and one event based model (DICHITOP). Thus, processing the precipitation ensemble with the set of hydrological models according to the above described method results in a sample of forecasted hydrographs $\mathbf{Q_\tau}$. The sample $\mathbf{Q_\tau}$ is statistically evaluated and provides characteristics concerning the range and reliability of flood forecasts as a function of τ.

Within this study, several past rainfall events comprising floods of different magnitudes are examined. Practical limits of forecast reliability and the use of information about forecast reliability from operational aspects are discussed with the operators of the early warning systems in the study basins.

Keywords: Flood forecasting, flood warning, ensemble forecast, multi model ensemble

Flood risk assessment in an Austrian municipality comprising the evaluation of effectiveness and efficiency of flood mitigation measures

C. Neuhold & H.-P. Nachtnebel
University of Natural Resources and Applied Life Sciences, Vienna, Austria

Even though substantial amounts were invested in flood mitigation in Europe the reported damages increased tremendously. One crucial reason was the frequently transformed land use in the former flood plains from agricultural utilization to industrial and residential areas. Obviously these modifications led to an increase of the damage potential.

The objective of this study was to analyse such a transformation process in detail on a case study in the city of Gleisdorf, in southern Austria. Several flood protection alternatives including structural and non structural measures were compared by estimating their effectiveness and efficiency. The alternatives referred to existing and conceivable flood mitigation measures along the river Raab. In this process the *structural measures* "dyke" and "retention basin" as well as the *non structural measures* "spatial planning" and "spillway" were evaluated by analysing historical data sets related to land use and coupled 1D-2D hydrodynamic modelling. A flood risk assessment based on a three-stage-methodology developed by BUWAL was conducted. Therefore the catchment of the river Raab was simulated by a semi-distributed precipitation-runoff model to provide the input data sets for the hydrodynamic model. Normative scenarios as well as logjam, dyke breach, sediment transport, vegetation, spillway and modified land use were considered. The simulated inundation lines, water depths and flow velocities were linked to the land use information to estimate the damage potential of the flood prone area. By integrating the scenarios and their respective probability a detriment was calculated. A valuable input to the assessment of the damage potential was delivered by the survey of numerous ex-ante and ex-post analyses for residential buildings and small trade as well as oral interviews with chief operation officers of local companies and decision makers. Dispatched questionnaires contributed negligible to the input data. The overall construction costs, the object related damage functions and the land use data provided the input for cost-effectiveness as well as benefit-cost analysis.

The results indicated that the effectiveness and efficiency of non structural and structural mitigation measures are within the same range. Negative effects on the damage potential and therefore a decreasing efficiency caused by land use change was clearly identified.

Keywords: Flood risk assessment, effectiveness, efficiency, structural measures, non structural measures, river Raab, Austria

EWASE—Early Warning Systems Efficiency – risk assessment and efficiency analysis

M. Gocht
Water & Finance, Berlin, Germany

K. Schröter & Manfred Ostrowski
Technische Universität Darmstadt, Darmstadt, Germany

C. Rubin
Pro Aqua GmbH, Aachen, Germany

H.P. Nachtnebel
Boku Vienna, Vienna, Austria

The EWASE project (Effectiveness and Efficiency of Early Warning Systems for flash-floods) is a R&D project within the ERA-NET CRUE integrated project. Two medium sized river basins prone to flash floods were considered, the Besòs basin (1020 km^2), Mediterranean climate, near Barcelona, Spain, and the Traisen basin (921 km^2), alpine climate, north east Austria.

The efficiency of an Early Warning System may be defined in two ways, firstly as the reliability of the forecast and secondly as the damage prevented by the forecast. Whereas EWASE elaborates on both issues, this contribution focuses on the second definition. EWASE pursues an innovative Approach in Risk Assessment and Efficiency Analysis. This approach can be ordered into three pillars: exposure analysis, susceptibility analysis and response analysis.

A major challenge for risk assessment was the international scope of the project. Traditionally international comparability of the results of risk assessment procedures has a low priority. But in EWASE the trans-European comparability was a key feature. Much emphasis in the vulnerability analysis was put onto exposure analysis which estimates the values of the assets at risk. Statistical data-sets sampled with an internationally standardised methodology were utilised. The important standards were the European System of Accounting ESA 1995 for macro economic accounting, the "Nomenclature générale des activités économiques dans les Communautés Européennes" (NACE) for activity classification, the Nomenclature of Territorial Units for Statistics, 2003 (NUTS 2003) for spatial distribution of asset values and regional as well as municipal employment statistics. Based on this sources capital intensity, defined as asset value per person employed, could be calculated. The product of a company's employees and its activity-specific capital intensity yielded an estimate of its asset values. A very conservative estimate was finally obtained by using (depreciated) net asset values.

In Susceptibility Analysis several sets of potential damage functions where used for calculating damages and giving a pragmatic estimate of associated uncertainty. Firstly a set well known in North-Rhine-Westphalia and developed for estimating the benefits of flood action plans was used. The internationally accepted approach for the ICPR Rhine Atlas was used as a second estimate. Thirdly for reflecting the properties of basins prone to flash floods results from the BMBF RIMAX project MEDIS where used, which focused on damage estimation in the ore mountains in Saxony based on data from the 2002 Elbe flood.

Response Analysis was based on a questionnaire based survey aimed at revealing the potential damage reduction through actions taken by the companies and residents in the basins. Finally, costs for implementation and operation of early warning systems were obtained from personal communication with suppliers and operators of such systems.

Results show that early warning bears a high potential of mitigating flood damages. The potential can be realised if local stakeholders and people at risk are well trained and prepared in interpreting early warning information. The more specific the information, the larger is the potential damage mitigation and the acceptance of the warning at the local level. Therefore long term stake-holder involvement is paramount to the efficiency and effectiveness of early warning systems.

Keywords: Flash flood, Risk Assessment, International Comparability

Flood risk management strategies in European Member States considering structural and non-structural measures

Jochen Schanze & Gerard Hutter
Leibniz-Institut für ökologische Raumentwicklung e.V., Dresden, Germany

Edmund Penning-Rowsell & Dennis Parker
Flood Hazard Research Centre (FHRC), London, UK

Hans-Peter Nachtnebel & Clemens Neuhold
University of Natural Resources and Applied Life Sciences, Vienna, Austria

Volker Meyer
Helmholtz Centre for Environmental Research—UFZ, Leipzig, Germany

Philipp Königer
Technische Universität München, Freising, Germany

Decisions about deploying structural (SM) and non-structural measures (NSM) for pre-flood risk management are made under manifold context conditions of decision makers. One of the context factors is the availability of appropriate evaluation possibilities to determine the complex and partly uncertain consequences of mitigation measures. Evaluation problems can particularly arise due to a lack of criteria, indicators, methods, knowledge and data. Since there is already some experience in evaluating SM, a research project has been carried out to derive a consistent methodology for evaluating both kinds of measures with an emphasis on NSM. The methodology is based on an evaluation framework with a set of criteria, indicators and methods. Major components are as follows:

- *Systematisation of structural and non-structural measures*: Based on previous classifications a suitable approach for consistent evaluation of SM and NSM is proposed.
- *Evaluation criteria*: Performance of SM and NSM may be evaluated regarding different aspects. The paper refers to effectiveness, cost-effectiveness, efficiency, sustainability, robustness, flexibility, and others.
- *Evaluation of non-structural measures*: Since there is little experience in evaluating NSM, indicators and methods are formulated to particularly assess effectiveness, cost-effectiveness and efficiency of such kind of measures. Thereby, also transaction costs are covered by the efficiency analysis.
- *Comparing SM and NSM*: Both kinds of measures can be complementary or alternative interventions in a flood risk system. While complementary effects may be analysed by specific criteria, indicators and methods for each single measure, comparison asks for similar aims, criteria and indicators. Methodical prerequisites of such comparisons are shown for some measures and criteria.

The methodology has been developed and tested in a number of case studies in United Kingdom, Scotland, Germany and Austria. The paper therefore presents detailed and site-specific results from evaluation. This for instance encompasses measures like building bans and the relocation of settlements and their comparison with structural flood defences also considering their performance under extreme events.

Generic outcomes are a comprehensive and consistent typology of flood risk reduction measures, the evaluation methodology and a framework of context conditions for a better understanding of decision making as part of flood risk management strategies. They are of a high relevance for the improvement of such of flood risk management strategies in Europe. Especially the integrative view on the evaluation capabilities and the context conditions of decision makers appears to be crucial for a better understanding and consideration of NSM.

The study is based on the ERA-NET CRUE project FLOOD-ERA.

Keywords: Flood risk management, measures, evaluation methodology, strategies, context conditions

Long term planning, integrated portfolios & spatial planning

The OpenMI-LIFE project – putting integrated modelling into practice in flood management

David Fortune
Wallingford Software, HR Wallingford Group, Wallingford, Oxfordshire, UK
Representing the OpenMI Association

Flood risk management is one highly specialized aspect of the wider field of water resource management. Some of the research and development intended to give broad benefits across a range of water management activities can certainly be used to good effect in flood management. One such result of research is the Open Modelling Interface, OpenMI. This paper describes progress in putting OpenMI into operational use for flood management and mitigation in the OpenMI-LIFE Environment project.

While the OpenMI is an undoubted technological success, its value will only be realized if integrated modelling itself is able to contribute widely to flood mitigation. And that depends on a dramatic shift in the way that institutions involved in flood management work together. In fact, it could be said that all too often they do not work together now.

The OpenMI-LIFE project's rationale lies in the Water Framework Directive, which demands an integrated approach to water management. This requires an ability to predict how catchment processes will interact. In most contexts, it is not feasible to build a single predictive model that adequately represents all the processes; therefore, a means of linking models of individual processes is required. This is met by the FP5 HarmonIT projects innovative and acclaimed solution, the OpenMI.

The purpose of OpenMI-LIFE is to transform the OpenMI from research output to sustainable operational product. It builds the capacity to use the OpenMI and demonstrates how it can be deployed, used, supported and funded at the operational level on real world scale problems. This demonstration is being conducted in co-operation with Competent Authorities in two Pilot River Basins, the Scheldt and the Pinios. It also shows how requests by users for changes to the interface are handled and implemented. It is the intention that the procedures and systems demonstrated and refined in this project will continue long after the project.

Four of the project use cases are directly relevant to flood management, as epitomized in Europe through the Urban Waste Water Treatment Directive and the Flood Risk Directive:

Scheldt Use Case A—The Impact of Sewer Discharges on a River during flooding.
Scheldt Use Case B—Interactions between downstream and upstream river flow regulations.
Pinios Use Case B—Impact of climate change scenarios on the reliability of a reservoir.
Scheldt Use Case D—Tides and upstream flood risk.

In this paper, these four use cases will be described. The project has been underway for over one year and the use cases have made varying degrees of progress. In particular, project participants have learned a lot about the difficulties we must all face in going from simplistic, single-model procedures towards new procedures based on integrated modelling. Technological change is not enough. Competent authorities, modelling specialists, and software suppliers will all have to change their attitudes and collaborate to a greater extent. This paper will describe the progress made and outline the lessons learned so far that should lead to institutional change in the future.

Keywords: Integrated modelling, institutional change, OpenMI

A method for developing long-term strategies for flood risk management

K.M. de Bruijn, M.J.P. Mens & F. Klijn
Deltares, Delft, The Netherlands

There are many different strategies to enable a region to cope with floods. Which ones are best, depends on the characteristics of the flood hazard, the region's vulnerability and the society's values and preferences. Since the characteristics of the flood hazard and the region's vulnerability change in time, flood risk management should be reconsidered regularly as well. Because the choice for a certain flood risk management strategy influences the options available for the future, a long-term view on the development of a region and the most suitable flood risk management strategy is required. This paper describes a method to develop and evaluate long-term strategies for flood risk management illustrated by an application on the Scheldt Estuary.

Long-term planning is challenging, because it involves thinking about possible future developments. In order to cope with the inherent uncertainty about this future, the proposed method advises the use of scenarios. Adopting a number of contrasting and clear scenarios reflects the large uncertainty associated with the future and illustrates the effects of a strategy in the context of different possible future situations. A scenario is defined here as all future autonomous developments, i.e. all future developments which do not purposefully influence flood risks. Strategies are defined as a coherent set of flood risk management measures.

The proposed method includes the following steps: system characterisation (area delineation, description and analysis of future developments), analysis of the current flood risk management strategy (current and future flood risk analysis and preliminary assessment), analysis of strategic alternatives (definition of strategic alternatives and analysis and preliminary risk assessment of the alternatives) and the full assessment of both the current strategy and strategic alternatives.

The method has been trialled on the Thames Estuary, the Elbe River and the Scheldt Estuary. This paper refers to the Scheldt Estuary application to illustrate the method. The studied part of the Scheldt Estuary consists of the Westerschelde water body and the surrounding low-lying polder areas. These polder areas are used for agriculture, housing and industry. Currently, they are protected by embankments which are designed to withstand once in 4000 year storm surge conditions. In the case study it is investigated what will happen with the flood risks if this strategy is continued into the future and what the effects of alternative strategies are. Not only flood risks are considered, but also socio-economic effects, effects on nature and effects on the system's sensitivity to uncertainties and future developments. The paper finally draws conclusions on long-term flood risk management for the Scheldt Estuary as well as on the usefulness and applicability of the long-term planning method.

Keywords: Flood risk management, long-term planning, scenarios, Scheldt Estuary, flood risks

The paper should be considered for the Young Floodsite Prize.

Flood Risk Mapping, using spatially based Systems Engineering

R. Raaijmakers
Grontmij Nederland b.v, De Bilt, The Netherlands

Flood risk management research has been developing intensively over the last decades. Flood risk management research data is, due to the development of the internet, widely available to the public, the scientific community and policy makers. Despite growing knowledge of both experts and lay-people, flood risk management will be progressively challenged in the following decades due to climatic changes.

The European directive on the Assessment and Management of Flood Risks entered into force on the 26th of November 2007 (European Commission, 2007). A principal part of the 'flood risk' directive is the composition of flood risk maps by each of the EU member states. Flood Risk Maps (FRMs) will display the possible negative consequences of probable flood scenarios. Each FRM will include a description of (a) receptors, (b) the economic consequences, and (c) installations, which are probable to cause accidental pollution that potentially affects drinking and bathing water and protected natural areas. In composing flood risk maps, each EU member state, has a degree of freedom of including (d) relevant information, additional to the items (a), (b) and (c).

Identification of all critical functions in a flood prone area, can be part of the development of effective flood risk management plans. Recent literature (Warm et al. 2007), promotes the identification and graphical representation of critical infrastructure, facilities or installations, fulfilling these critical functions, in FRMs. A hospital is an example of a critical facility.

The susceptibility of an object (eg. tangible or intangible assets), reflects its sensitivity to a natural hazard (Green, 2004). For critical objects, susceptibility of the object itself, influences the magnitude of the direct and indirect consequences of flooding. This means critical objects (1) fulfil a vital function in operational flood risk management; or (2) the function of the object in the pre-flood stage, increases the vulnerability of a case study area to the consequences of flooding, during a flood and in the post-flood stage. Within this paper critical objects are identified, using a Spatial Systems Engineering (SSE) approach, building upon the approach of Everleigh (2006).

Our SSE approach identifies the critical functions during a flood within the spatial boundaries of a case study area. The case study area covers the inundated area itself and the area affected by the primary indirect effects of flooding (Smith & Ward, 1998), for a flood scenario. Critical functions are attributed to critical objects within the case study area. Subsequently, the position of each object and the (spatial) correlation between the critical objects is determined. The inundation characteristics of the flood scenario are used to obtain the effect of a flood onto the functioning of the critical object. Thereafter the effects caused by the (partial) failure of objects within the case study area onto socio-economical and intangible floodplain functions and values (Blackwell and Maltby, 2006) are identified.

Finally, all obtained data is integrated and visualized within a FRM. On the basis of this map, the functional requirements of the critical objects are determined, both in the basic pre-flood situation, as well during a flood. Concluding, alternative positions and additional functional requirements for the critical objects, are analysed. This study gives a practical insight of the interdependency of asset management and flood risk management strategies.

Keywords: Flood Risk Maps, Spatial Systems Engineering, disaster management

Finding a long term solution to flooding in Oxford: The challenges faced

L.G.A. Ball
Black & Veatch Ltd., Redhill, Surrey, UK

M.J. Clegg
Black & Veatch Ltd., Chester, Cheshire, UK

L. Lewis
Environment Agency, South East Area Office, Frimley, Surrey, UK

G. Bell
Environment Agency, West Area Office, Wallingford, Oxon, UK

This paper investigates the challenges that face the Oxford Flood Risk Management study team and examines how they are being addressed in and incorporated into the development of a long term Strategy for Oxford.

Flooding is an ever-present threat for those parts of Oxford and the surrounding villages that lie on the converging floodplains of the River Thames and the River Cherwell. Serious flooding has occurred in Oxford three times in recent years, in December 2000, January 2003, and most recently last summer. The return period of these events were each between 1 in 10 and 1 in 20 years with as many as 150 properties being flooded internally and major disruption to all aspects of city life. The worst affected areas were Osney, Botley, New Hinksey and South Hinksey.

A 1 in 100 year flood event would be far more widespread and would threaten 3,500 residential and commercial properties; schools, colleges and health centres; scheduled monuments and other heritage sites; key transport links and substantial areas of recreational, amenity and agricultural land.

To reduce this risk, the Environment Agency has commissioned consultants Black & Veatch Ltd to work with them to develop a long term (100 year) Flood Risk Management Strategy for Oxford which recognises the impacts of climate change. To achieve this goal many technical, environmental, legislative and economic issues must be addressed. This paper investigates how these issues complicate the process of identifying a sustainable long term strategy.

The complex nature of how Oxford floods presents the greatest technical challenge. The most obvious source of flooding is from out-of-bank flows from the Thames and the intricate network of braided streams that flow throughout the western part of the city. For many areas of the city, however, the first indication of imminent flooding is rising groundwater, which pushes up from the gravels and through the thin layer of alluvial clay that lies over much of the floodplain. Man-made features such road and rail links and land fill sites also act as barriers to the movement of floodwaters, and surface water drainage systems complicate the picture further.

The environmental interest of the study area is also significant and just as complex. Oxford boasts a wealth of nationally and internationally important ecological, heritage and landscape features. These features are spread throughout the flood risk area, but are perhaps best typified by Port Meadow: a site designated under EU and UK law as a Special Area of Conservation, a Scheduled Monument and recognised as part of the 'Landscape of Key Significance' by the City council. Critically this site, like other important conservation sites within Oxford, is dependent on annual flooding to keep them special. To stop this type of flooding would lead to unacceptable impacts and change the character of Oxford itself. The Strategy will need to be developed with this in mind.

To deliver a flood risk management strategy for Oxford, the Environment Agency must show that it provides the best value to the UK economy. The Strategy will need to demonstrate that the money saved by reducing the risk of flooding far outweighs the money required to implement it. In addition, it will need to compete for a limited national pot of funding from the Government with other towns and cities that are also badly affected by flooding.

Keywords: Flooding, strategy, flood risk management

Risk analysis and decision-making for optimal flood protection level in urban river management

M. Morita
Shibaura Institute of Technology, Tokyo, Japan

Urban river basins with rapidly concentrated population and assets have high flood damage potential in both developed and developing countries. To prevent flood inundations, national and local governments are obliged to undertake and carry out structural flood control projects along with non-structural measures. The available financial resources, however, would be limited for the implementation of any proposed projects. Effective and efficient projects are, therefore, widely emphasized among decision-makers, the municipal engineers in charge of infrastructures for flood control management. In particular, for flood control planning, a reasonable and appropriate flood protection level is naturally demanded in terms of the budget capacity of the governments and the expected security of the people living in river basins. Engineering methods with the reasonableness based on risk analysis should be established to support the decision-making process for an optimal flood protection level.

The objective of the study is to present a methodology of risk analysis for optimal flood protection level decision in the framework of flood risk management for urban river basins. Risk analysis for flood risk management schemes is already a widespread practice for flood control planning. The study, however, deals with urban river flood control where the flood protection level is decided based on intensity-duration-frequency curves having their own return periods. The concept of "risk" is here defined as the product of flood damage potential and its occurrence probability.

For the risk analysis a flood damage prediction model **FDPM** using GIS is applied to calculate flood damages for any design storms with different return periods. **FDPM** is composed of two models: a flood inundation prediction model (**Model 1**) and a flood damage estimation model (**Model 2**). In the study GIS with private and corporation assets data of the river basin is effectively utilized for flood damage calculation by **Model 2** to overlay the assets data with the inundation depths calculated by **Model 1**. The calculated monetary damages for the design storms with their occurrence probabilities enable us to quantify flood risk as an **Annual Risk Density Curve** based on the concept of "risk." The annual risk density curve gives the relation between design storm level and its flood risk density. The expected value obtained by the integral of the risk density curve generates the annual flood risk cost.

FDPM and the risk analysis method stated above were applied to the optimal flood protection level decision of the Kanda river basin in the Tokyo Metropolis, Japan. The river basin, having an area of 80.6 km^2, is highly concentrated with population and assets.

If a measure for flood control is taken in order to make the river basin safer with respect to inundation, the annual risk cost becomes lower and consequently a corresponding capital cost is needed for the flood protection. As a flood protection level or an adopted return period for flood control becomes higher, the capital cost increases, and the annual risk cost reduction increases owing to the better protection afforded. Whereas the annual risk cost reduction is in the study shown as a **Risk Cost Reduction Curve**, the corresponding capital cost is given as a **Capital Cost Curve**. The optimal flood protection level can be determined by benefit-cost analysis using the risk cost reduction curve and the capital cost curve. The study deals with two cases as a flood control measure: a flood control reservoir and storm infiltration facilities. The flood risk analysis shows how to decide an optimal flood protection level and discusses what is "optimal" in terms of effective performance of capital cost, total cost reduction, and available financial resources.

Keywords: Flood risk analysis, decision support, optimal flood protection level

An integrated risk-based multi criteria decision-support system for flood protection measures in riversheds—REISE

Nils P. Huber, Daniel Bachmann, Holger Schüttrumpf & Jürgen Köngeter
Institute of Hydraulic Engineering and Water Resources Management, RWTH Aachen University, Germany

Uwe Petry, Markus Pahlow & Andreas H. Schumann
Institute of Hydrology, Water Resources Management and Environmental Engineering, Germany

Julia Bless & Gottfried Lennartz
Research Institute for Ecosystem Analysis and Assessment, RWTH Aachen University, Germany

Oliver Arránz-Becker & Manfred Romich
Institute for Sociology, RWTH Aachen University, Germany

Jürgen Fries
Wupperverband, Germany

INTRODUCTION

River flood protection measures are nowadays planned and optimized on the basis of risk. Flood risk analyses are supported by the availability of elaborate engineering methods and tools. These allow for the detailed analysis of hydrologic and hydraulic processes, flood recurrence intervals and socio-economic damage. As a result, flood risk management activities in Germany are mainly undertaken on the basis of scenario-based approaches for bounded river stretches showing large potential damage. Risk-related feedbacks between flood protection measures provided at different locations in the rivershed are usually not regarded. This can only be accounted for within an integrated risk management for a complete rivershed. As in the correct mathematical sense integral flood risk for a rivershed comprises the unbounded set of possible flood events, potential flood damage and also the reliability of flood protection measures, new risk management concepts and methods have to be developed.

APPROACH

The scope of the proposed paper is to present concept, methodological challenges and appropriate solutions for supporting decision-making in the field of optimum flood protection on the rivershed scale. The interdisciplinary project REISE, funded within the german RIMAX-initiative, focuses on the development of a decision support system (DSS) for the optimization of protection measures against extreme flood events in small to medium riversheds. The project structure is shown in figure 1.

Hydrological analyses and simulations result in the characterization of precipitation and runoff ranging from frequently occurring to extreme events. These may cause failure of technical or organizational flood protection measures which is accounted for probabilistically in the DSS. Potential damage in the rivershed is linked to natural inundation or breach-induced flooding. A special focus is set on the consideration of potential damage from the correct interdisciplinary point of view. Tools for deriving the overall damage and also costs of measures, i.e. economic, ecologic and psycho-social, on an uniform level of detail and significance are developed. Risk and costs associated with these specific fields have to be merged to one statement. This multi-criteria approach is a key challenge within the DSS.

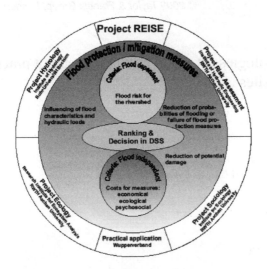

Figure 1. Concept of the project REISE.

Special attention is also payed to the demand for a methodically elaborate but nevertheless still manageable and efficient DSS. A major issue is the computation time for representing all possible system states in a rivershed on a meaningful level. Therefore, the way of analyzing the total risk space for a rivershed, including the multitude of possible performance-states of flood protection measures, will also be specified in the paper.

Keywords: Risk management, integrated decision making, extreme floods, rivershed scale, optimization

Integrated methodologies for flood risk management practice in European pilot sites

Jochen Schanze
Leibniz-Institut für ökologische Raumentwicklung e.V., Dresden, Germany

Peter Bakonyi
Water Resources Research Centre (VITUKI), Budapest, Hungary

Marco Borga
University of Padova, AGRIPOLIS, Legnaro, Italy

Ben Gouldby
HR Wallingford, Wallingford, UK

Marcel Marchand
Deltares, Delft, The Netherlands

José A. Jiménez
Universitat Politécnica de Catalunya, Barcelona, Spain

Horst Sterr
Geographisches Institut UAG Küstengeographie & Naturgefahren, Universität Kiel, Kiel, Germany

Flood risk management is a societal challenge: It encompasses the analysis of risk generation in complex human-environment systems, the evaluation of the tolerability of risk as well as the design and implementation of risk reduction measures. Multiple actors are involved in the process of collecting information, decision making and acting. Science may support this endeavour by specific natural and social science investigations and software tool developments. The complex levels of flood risk management, however, require a comprehensive understanding of the entire issues and special methodologies for the emergent tasks.

The paper presents results from developing and testing integrated methodologies to support flood risk management in European pilot sites of the FLOODsite project. Based on a conceptual framework of the societal flood risk management it illustrates how significant advances can be made by context-specific integrated approaches. The framework describes flood risk management from the perspective of multiple decision makers and experts who are responsible for dealing with flood risk as a real-world problem and related management tasks. Moreover, it explicates the management process itself considering the context and process dimension in addition to the contents.

Following this framework, integrated methodologies and their contributions to flood risk management practice are shown from seven European pilot sites, the Elbe River basin, a network of European flash flood basins, the Tisza River basin, the Thames River estuary, the Scheldt River estuary, the Ebro River Delta coast and the German Bight coast. With respect to the first task of risk management, which is *risk analysis*, it is shown how coupled modelling of flood hazards and flood vulnerabilities can improve the societal understanding of riverine, estuarine and coastal flood risk systems and their dynamic. Major features of the different approaches are detailed, as well as, their potential role in supporting the implementation of the Floods Directive.

For the second task of risk management, *risk evaluation*, an overview is given on findings regarding appropriate criteria and methods. This covers both the ex post and ex ante risk evaluation of the flood risk system. It also deals with the performance of *risk reductions* options, which links to the third task of risk management. In this respect, the meaning of certain physical measures and policy instruments for the pilot sites is explained and possible improvements for selected measures are derived.

For the interface between these three tasks and the *flood risk management process,* decision support tools play an increasing role. Different software tools have been developed and applied at three pilot sites. Principle differences between these tools are explained and briefly discussed in the light of their functionality and societal applicability. Based on this, results from site-specific investigations on risk perception, strategic planning and stakeholder involvement lead to the initial presumption that science needs also to address the social challenge of flood risk management. Investigations suggest significant differences between the requirements of the various actors of flood risk management and the people at risk. Final conclusions are thus drawn, with respect to the future elaboration of flood risk management plans and public involvement according to the Floods Directive.

Keywords: Flood risk management, methodologies, risk analysis, risk evaluation, risk reduction, pilot sites

Underpinning flood risk management: A digital terrain model for the 21st century

Mark Stileman & David Henderson
Ordnance Survey, Southampton, UK

Modelling the surface of the landscape has always been one of Ordnance Survey's core activities. During the early 1990s, its paper repository of 1:10,000 scale contours was digitised under an intensive 5-year programme and Great Britain's first national digital terrain model (DTM) product, Landform PROFILE, was subsequently created. Since this time, user requirements have become considerably more demanding. Since 2006, as part of the Atlantis[1] initiative, Ordnance Survey has committed to deliver a fit-for-purpose DTM which is interoperable with other core reference data, for example the Environment Agency's Detailed River Network, in fulfilling the needs of legislative drivers such as the Water Framework Directive and INSPIRE.

In 2005 Ordnance Survey released an enhanced DTM, Landform PROFILE Plus, as a first step towards the realisation of terrain data more suited for 21st century applications related to water and flood management. A key facet of this has been explicitly engineering the data to account for landform features and man-made structures which significantly affect the movement of water in the landscape. The priorities for a national DTM today can be summarised as *variable specifications, product interoperability* and *change-driven maintenance*.

Populating a national DTM requires robust data collection methods. The role of remote sensing has changed dramatically over the last two decades. Advances in technological capability are manifest, with traditional photogrammetry being enhanced by increasing automation and improved levels of accuracy through all-digital flowlines, and lidar methods demonstrating rapid evolution with improved sensors and new waveform capabilities. Ordnance Survey is actively researching new techniques and processes for capturing terrain data with increasing effectiveness and efficiency.

The landscape of Britain is dynamic. This needs to be reflected in the appropriate maintenance of DTMs. This has been problematic in the past because entirely replacing terrain data for a given area can be expensive where there has been limited change. In the short term, improved change intelligence processes can be deployed to target revision in a smarter fashion and additionally there is scope to more tightly integrate historically disparate data collection flowlines. In the longer term, there is an exciting opportunity to integrate terrain data more fundamentally with Ordnance Survey's other data core data within a database environment. In addition to considerably improving the interoperability of product types, this would also help to revolutionise the way that DTMs are maintained.

As the national mapping agency, Ordnance Survey's primary role is to provide the underpinning geographic referencing framework for Great Britain. Its DTM strategy is to fulfil this by improving the relevance of the content as well as moving closer towards a vision of truly integrating a national DTM with topography, address and transport data, interoperable with externally-generated geographic reference datasets, thus enabling a far more joined-up approach to flood management. This paper discusses how, through a combination of strategic planning and the judicious adoption of new technology, this vision will be delivered.

Keywords: Digital terrain model, DTM, national, variable specification, accuracy, resolution, change, maintenance, integration, interoperability

---

[1] Atlantis website—http://www.dnf.org/applications/atlantis

Integrated land and water management in floodplains in England

H. Posthumus
Cranfield University, UK

J.R. Rouquette
Open University, UK

J. Morris & T.M. Hess
Cranfield University, UK

D.J. Gowing
Open University, UK

Q.L. Dawson
Cranfield University, UK

Agricultural Flood Defence Schemes in floodplain and coastal areas were once an important element of Government support for farmers in Britain. More recently, however, changing priorities in the countryside, concern about environmental quality and perceptions of increased flood risk in lowland areas, in part linked to climate change, have promoted a re-appraisal of land management options and policies for floodplain areas.

With support from the UK Economic and Social Sciences Research Council*, a number of Agricultural Flood Defence Schemes in England, previously studied by the research team in the 1980s, are being re-examined to identify and explain changes in land and water management that have occurred over the last 40-years. Methods include stakeholder and institutional analysis, farmer interviews, field observations and modelling of hydrological and related ecological processes. The influence of agricultural policy, interacting with farmer circumstances and motivation, is also explored.

The paper explains the application of a conceptual framework (Figure 1) that integrates the main components of the research. It uses the concept of ecosystem functions and services as a basis for representing the major

Figure 1. Conceptual framework for integrated floodplain management.

---

* as part of the Rural Economy and Land Use (RELU) Programme.

domains of flood risk management, farming systems and ecology, as well as linking these aspects to dominant stakeholder interests. The paper reports the use of generic scenarios developed for each study site to consider management options that focus on single objectives, such as maximising agricultural production, maximising biodiversity and minimising flood risk to built development in the catchment. It also explores options which attempt to combine multiple objectives. This emerging analysis demonstrates how an integrated, ecosystems approach can helps to inform future policy and practice for floodplain management, hopefully in ways that appeal to key stakeholders.

Keywords: Ecosystem services, floodplain management, hydrology, ecology, rural economy

Putting people and places at the centre: Improving institutional and social responses to flooding

C. Twigger-Ross & A. Fernandez-Bilbao
Collingwood Environmental Planning, London, UK

L. Colbourne
Lindsey Colbourne Associates

S. Tapsell
Flood Hazard Research Centre, Middlesex University, UK

N. Watson, E. Kashefi, G. Walker & W. Medd
Department of Geography, Lancaster Environment Centre, Lancaster University, UK

Flood risk management is at a crossroads. Recent flood policy for England and Wales (Making Space for Water, 2005), events in the North East and South West of England (Summer, 2007) and their reviews (e.g Pitt Review, 2007) have converged around the need to move from flood defence towards flood risk management. This shift requires an approach to flooding that puts people and places at the centre of policy and practice. This will require strong leadership by those in management positions. To support this change the institutional and social processes integral to flood and coastal erosion risk management have to be understood and developed. This research focussed on an investigation of the social and institutional aspects of flood warnings, response and resilience, community engagement and institutional culture. Four key questions were examined within the project:

How to move the current flood warning service delivered by the Environment Agency, England and Wales from a warning based service towards a response based service?

How to understand and develop the role of the Environment Agency, England and Wales in building resilience within communities?

How to embed appropriate collaboration and engagement within flood incident management in the Environment Agency, England and Wales?

How to develop an agreed, consistent institutional approach to flood incident management within the Environment Agency, England and Wales?

The project used a combination of methods to review information and collect data: standard research review, questionnaires and interviews with Environment Agency flood incident management staff, a case study focussed on one place two years after a flood that used focus groups and interviews to explore key issues, together with the development and testing of practical tools for collaboration and engagement post-floods.

From the four strands of research a final synthesis has been carried out and this paper focuses on that synthesis and the recommendations from the whole project. The recommendations have been developed as scenarios for change, providing practical advice for each aspect addressed by the project as to how that change might be implemented, and an assessment of the benefits and limitations to each scenario.

Keywords: flood incident management, social science, engagement, collaboration, England and Wales

Delivering Integrated Urban Drainage – current obstacles and a proposed SUDS planning support tool

V.R. Stovin, S.L Moore & S.H. Doncaster
Pennine Water Group, Department of Civil and Structural Engineering, University of Sheffield, Sheffield, UK

B. Morrow
United Utilities, Warrington, UK

Traditional approaches to surface water drainage in urban areas rely on piped sewer systems to remove the runoff as rapidly as possible. Piped systems (separate or combined sewers) tend to increase the speed and volume of surface runoff, representing a significant element of the flood risk in many situations. These systems also have detrimental impacts on urban water quality, and fail to recognise the resource value of water. Increasing urbanisation and climate change are expected to worsen the impacts associated with stormwater runoff. Retaining and attenuating stormwater on-site through the use of sustainable drainage systems (SUDS) will assist in reducing flood risk in addition to improvements in water quality. SUDS may also promote groundwater recharge and provide benefits for the wider social and ecological environments. In this context, the renaissance in urban living presents exciting opportunities to reconsider surface water management within the urban landscape.

The paper focuses on The Lower Irwell IUD project, which is centred on the River Irwell floodplain in Salford. The project is one of fifteen Integrated Urban Drainage (IUD) pilot projects instigated by Defra during 2007. The area has a history of flooding associated with multiple causes. The project has considered the interaction of stakeholder organisations with respect to the development of strategies and delivery mechanisms for IUD. The paper will summarise key limitations and constraints identified within current planning and regulatory systems that prevent an integrated approach to urban drainage. Key stakeholders include the local authority, water utility, the Environment Agency and developers, in addition to national organisations including Defra, OFWAT and the Association of British Insurers. The main constraints that were identified as preventing a more integrated approach to urban water management include: data and software incompatibility; data confidentiality, ownership and liability; resources; and limitations in local knowledge and expertise. Regulatory constraints provide virtually no incentive for SUDS approaches to be implemented by any of the key stakeholders.

The IUD pilot project focuses on an area of social deprivation and low income, and the City Council is overseeing a significant programme of urban regeneration and renewal. Planning on this scale provides a unique opportunity to implement (retrofit) SUDS into urban areas. It is, however, vital that a clear and consensual view on what is required is understood by all the relevant stakeholders and that this underpins and negotiations that take place between planners and developers. Local authority planners have limited capacity to persuade developers to consider SUDS within new developments, including regeneration projects. This paper will outline a prototype SUDSMap, aimed at providing planners with simplified guidance on where SUDS technologies should be sought, and which types of SUDS might be most appropriate. The authors envisage that the SUDSMap might be developed in parallel to the Local Development Framework. The SUDSMap encapsulates topographical, geological and groundwater constraints, in addition to the planning opportunities associated with redevelopment. In the prototype SUDSMap, zones are delineated as "appropriate for infiltration SUDS" or "not suitable for infiltration, use green roof or porous pavement", for example. The SUDSMap will represent the stakeholders' consensus, and is seen as a key part of any IUD strategy.

Keywords: Urban drainage, SUDS, flooding, runoff, planning

Strategic planning for long-term Flood Risk Management – findings from case studies in Dresden and London

G. Hutter
Leibniz Institute of Ecological and Regional Development (IOER), Dresden, Germany

L. McFadden
Flood Hazard Research Centre (FHRC), Middlesex University, London, UK

Moving from defensive towards pro-active strategies is becoming widely recognised in Flood Risk Management (FRM). Pro-active strategies involve strategic planning to think and act ahead of the build up of risk. However, not all challenges faced by decision-makers in moving towards this ideal are well-known. This paper highlights challenges of strategic planning as political process in Dresden and London. Concepts and empirical findings are results of FLOOD*site*-Task 13.

FRM emphasises that managing flood risk is a holistic, continuous, and adaptive process. In line with this, four characteristics were considered as indicative of strategic planning:

- Nurturing diversity in decision-making,
- Embracing uncertainty and change,
- Combining the range of existing knowledge systems into the decision-making process, and
- Creating opportunity for self-organisation to generate new knowledge.

The *Dresden* study explored complexities involved in managing flood risk after the Weisseritz flash flood disaster in August 2002. The study shows that it is difficult for local and state decision-makers to *maintain* diversity and to embrace uncertainty after the flood disaster. At first, the disaster triggered attempts to increase local diversity in measures (e.g., spatial planning for controlling economic assets on flood plains). Then, decision-makers focused on restoring order and on the pre-disaster strategy of defending against floods through structural measures (e.g., building new flood walls). The study shows that this is due partly to difficulties in combining existing knowledge systems in *local* government and administration. It concludes that strategic planning requires both: an organised approach to strategic planning with different roles and relationships (e.g., sponsor, process-champion) as well as self-organisation (e. g., forum for communicating about residual risk).

The *London* study highlights a series of important steps that have been taken towards strategic planning in the London and Thames Estuary region. However, the fact remains that an integrated and adaptive approach to planning remains difficult to achieve. Ensuring all stakeholders are engaged is critical to understanding the range of conflicts and choices. Yet, a transparent, accountable and just decision-making process demands a radically different approach to existing models of planning and implementation. Nurturing diversity requires difficult questions concerning the distribution of power and funds to be addressed and in a large urban area such as London the political nature of decision-making can become greatly magnified. Furthermore, developing a planning process based on a continuous cycle of learning and testing among stakeholders is an important challenge. However, learning from previous flooding events and the ability to self-organise is often constrained within stakeholder organisations: resources, perceptions and politics being three primary limiting factors.

The paper concludes that the conceptual ideas of strategic planning for long-term FRM are easy to understand, but difficult to implement. Difficulties arise because strategic planning it not only a technical and science-driven process based on a common "language" and common interests. Strategic planning is also a political process based partly on different perceptions (e.g., between local flood risk managers and planners) and diverging interests (e.g., embracing uncertainty vs. restoring order). As a consequence, understanding and steering governance becomes important for FRM.

Keywords: Strategic planning, adaptive management, long-term flood risk management, risk governance

Extreme flood events & flood management strategy at the Slovak-Austrian part of the Morava river basin

M. Lukac & K. Holubova
Water Research Institute, Bratislava, Slovakia

Morava river is a left sided tributary of the Danube river. Its river basin is located in three countries—Czech Republic, Austria and Slovak Republic. The river basin suffered from large floods during the last decade. The consequences of 1997 flood were catastrophic in the Czech part of the basin, while those of 2006 flood mainly in Austria. These floods pointed at the weaknesses in the flood management and active preparedness during similar extreme events. Consequently, great effort has been put into the design of alternative strategies and measures for minimising flood risk. Flood damages and experiences initiated several local, regional, as well as international projects aimed at the improvement of flood management in this territory. Some of these projects also include the aims of improvement of unique wetland ecosystem of the Morava river in order to restore natural river functions.

Implementation of recently approved European Directive on the assessment and management of flood risk requires harmonization of flood risk management plans in the international river basins. To agree on the common strategy of flood management improvement is the effort of competent authorities at both sides of the lower Morava river. Results obtained from several Slovak-Austrian EU projects, case studies and research projects focused at flood management issues are presented in the paper. The emphasis is put at the evaluation of the current state, study of the effectiveness of various flood retention measures, as well as provision of tools for decision support in the crisis situations. Hydrodynamic numerical modelling and GIS tools created the core of applied solutions.

Results presented in the paper will show the effectiveness of flood control measures (polders, enlargement of floodplains by shifting the flood protection dykes) proposed by Austrian side, as well as integrated effect of all polders proposed at both sides of the Morava river floodplain. Based on the results, individual scenarios are compared and consequently optimum parameters of flood control measures are defined. Scenarios assuming integrated effect of all polders is evaluated as the best one regarding enhancement of flood protection (decrease of flood water level, routing of floods) as well as restoration effect in the frame of previous flood plain. Some further possibilities for increasing of effectiveness of flood control measures are indicated.

Flood hazard maps created for dyke breaks are evaluated for the most endangered areas at both sides of the river. The maps of simulated water depth in the flooded area represent valuable information, which can be utilized in the decision support and crisis management in the hazard zones and local communities.

Presented results also document very good and effective cooperation between Slovak and Austrian experts to cope with common problems of flood protection on the international territory of the Morava river.

Keywords: Flood management, international river basin, numerical modelling, decision support

… *Flood Risk Management: Research and Practice – Samuels et al. (eds)*
© *2009 Taylor & Francis Group, London, ISBN 978-0-415-48507-4*

Using non-structural responses to better manage flood risk in Glasgow

Richard Ashley & Richard Newman
Pennine Water Group, University of Sheffield, UK

Fiona McTaggart
Scottish and Northern Ireland Forum for Environmental Research, UK

Sandy Gillon
Glasgow City Council, UK

Adrian Cashman
University of the West Indies, Barbados, WI

Gina Martin
Scottish and Northern Ireland Forum for Environmental Research, UK

Susan Molyneux-Hodgson
Department of Social Studies, University of Sheffield, UK

Traditional structural approaches to managing flood risk are expected to be too expensive to be employed to counteract all of the effects of climate change in the UK[1] or even to address all of the risks from recent flooding events. In view of this, and similar findings elsewhere in Europe, a project has been underway as part of the EU CRUE ERA-NET[2] programme that aims to address Risk Assessment and Risk management: Effectiveness and Efficiency of Non-structural Flood Risk management Measures in small urban catchments. This paper deals mainly with the activities undertaken in the Glasgow case study which set out to review and identify 'best' practice from Glasgow for 'non-structural' measures in terms of delivering greater resilience to flooding.

In 2002, Glasgow experienced the worst flooding in living memory, from rivers, small watercourses drains and sewers as well as local pluvial sources. To address this, the Scottish Government and main stakeholders have developed the Glasgow Strategic Drainage Plan (GSDP), based on detailed modelling studies to identify causes and potential responses. In addition, strong linkages, networks, partnerships and collaboration between the main stakeholders have established a new and coordinated approach to integrated flood risk management in Glasgow (and Scotland) that is based on a number of Flood Liaison Advisory Groups (FLAGs). Scots Law[3] has set out an approach to flood risk management (FRM) that is based on sustainability. What this means in practice has been formulated via the National Technical Advisory Group (NTAG) to mean: '..*the maximum possible social and economic resilience against flooding, by protecting and working with the environment, in a way which is fair and affordable both now and in the future.*' Application of this in practice has now been taken up by the Flooding Issues Advisory Committee (FIAC)[4] and includes a definition of 'resilience' that encompasses what is termed 'the four A's': Awareness of the risks of flooding; Alleviation of flood effects; Avoidance of flood risk if possible; and Assistance in the event of a flood.

The case study in Glasgow utilised face to face interviews with main stakeholders; with the public and community groups; group meetings and discussions in order to identify the opportunities for and barriers to the use of non-structural measures (NSMs) alongside appropriate structural measures. It investigated application of NSMs in a catchment in the East End of Glasgow and looked retrospectively at their possible application

---

[1] Foresight future flooding report (2004).
[2] http://www.crue-eranet.net/
[3] http://www.crue-eranet.net/
[4] Sustainable Flood Management. FIAC. FIAC2006(6).

in already completed projects. It is concluded that the main barrier to uptake of these measures is a lack of experience and knowledge of their performance, costs, acceptability and hence overall sustainability, although institutional and regulatory barriers also militate their use. The project set out also to develop the capacity in the various stakeholder groups to consider and utilise these measures and also to facilitate the transition from 'technological entrapment' evident in current approaches, through capacity building to the use of these and other measures that can lead to more resilient flood risk management. The paper outlines the processes used in the study and provides new insights into the selection of non-structural measures based on a GIS system depicting vulnerability, capacity needs and strengths and the relative performance of non-structural measures in terms of FRM (economic, technical and social).

Keywords: Flood risk management, non-structural responses, capacity building, European practice

Vulnerability and resilience, human and social impacts

The policy preferences of citizens, scientists and policy makers

J.H. Slinger & M. Cuppen
Delft University of Technology, Delft, The Netherlands

M. Marchand
Deltares, Delft, The Netherlands

The knowledge of flooding, the perceptions of flood risk and the policy preferences of citizens living alongside the transnational Scheldt Estuary, scientists studying the system and local and regional policy makers are described in this paper. Of particular interest are the policy preferences of these three types of actors following the communication of new insights from flood modelling studies. As such, we address the role of new knowledge of flooding and the opinions of fellow citizens in influencing the preferences of actors regarding flood risk management options.

This study was conducted within the context of the EU-funded Sixth Framework Programme FLOODsite and forms a component of the Scheldt Estuary Pilot Study. In 2005, semi-structured interviews were conducted with local residents on either side of the Dutch-Belgian border. Similarities and differences between people of similar profession and age yet different nationality were identified. Insights from this analysis were subsequently communicated to policy makers at an international workshop and substantive information was cross-checked. Policy makers were surprised at the accuracy of the information supplied by the local residents and the percipience of their judgements regarding the potential consequences of flooding. They expressed a desire to be informed of the progress of the study and a willingness to participate in a final workshop three years later.

In the interim, flood modelling studies were conducted and a scenario-based analysis of potential future flooding risks was undertaken (de Bruijn et al. 2007). This information was communicated to local citizens by the involved scientists at a subsequent workshop after all present had received information on the analysis of the original interviews. The policy preferences of both groups prior to the workshop and after the exchange of information were assessed on the basis of a questionnaire in which they were requested to allocate priorities to various flood risk management measures. These ranged from primary defences such as dikes to evacuation and post-flood compensation. Both the scientists and the local citizens placed a high priority on the primary defences in the form of dykes. The farmers were opposed to large-scale managed realignment of the dykes (depoldering) and were later joined by the scientists. A noticeable effect was that all participants spread their priorities further along the flood hazard chain by including post-event management measures following the workshop.

In the final workshop with policy makers, the new model-based information as well as the new insights on the policy preferences of citizens and scientists were communicated. The policy preferences of the policy makers prior to and following the workshop were assessed using the same questionnaire and the influence of new knowledge on the policy preferences of the policy makers concerned with the Scheldt Estuary was established. Unlike the situation with the local citizens, strong differences in policy preferences were found between Dutch and Belgian policy makers. However, the policy makers did evince a trend towards flood risk management rather than flood defence alone and this concurred with the reactions of the citizens and scientists.

In summary, this study successfully examines the role of new knowledge in informing the policy debate amongst different actor groupings in the complex socio-technical arena termed flood risk management.

Keywords: Socio-technical system, flood modelling, policy studies, flood risk management, stakeholders, public involvement

Analysis of the human and social impacts of flooding in Carlisle 2005 and Hull 2007

P. Hendy
National Flood Forum, Bewdley, Worcestershire, UK

No one who has witnessed a disaster is untouched by it. The reality is that for the flood victim the recovery stage is often worse than the disaster itself. The impact of the event affects not just the property he lives in but also has repercussions on his family, health and work situations.

This study is drawn from the flood recovery programme in Carlisle (2005) and Kingston upon Hull (2007) and details many of the lessons learned. It focuses on the often ignored effects of flood trauma on the individual. This also has important implications on the local Community and its ability to delay or promote recovery.

Much of the analysis is drawn from detailed involvement with and feedback from flood victims themselves, as they faced the task of rebuilding their lives.

- Following flooding the victim is often relocated to alternative accommodation. The analysis and comparison that was conducted showed that what housing was provided by the Insurance Industry had a major impact on people's health and ability to cope with the event.
- One major failure in Disasters/Disaster recovery is the failure to communicate effectively with the victim. Whilst many Insurance companies prided themselves on lessons learned from the 'Carlisle Experience' the reality was very different in Hull.
- What became apparent was that the additional trauma on flood victims seen in Carlisle as Christmas approached (12 Months after the flood)—also became apparent in Hull over the Christmas period (6 Months after the flood).

This paper also details the work that was undertaken by the Voluntary Sector, where the executive officers from the charitable sector were faced with the dilemma of supporting the community/individuals whilst at the same time facing mounting costs in providing a necessary service. The phrase often heard was 'used and abused'.

An issue in developing flood/disaster recovery is understanding the length of time that is required to 'normalise' a community. Through the analysis and comparisons of two very different communities the recovery model that was established seeks to ensure that the support and help that is needed is provided in the most effective way.

Keywords: Flood Recovery, Health and Social Impact on Flooded Communities, Trauma in Flooded Communities, Flood Insurance Issues

Institutional and social responses to flooding from a resilience perspective

N. Watson, E. Kashefi, W. Medd & G. Walker
Lancaster Environment Centre, Lancaster University, Lancaster, UK

S. Tapsell
Flood Hazard Research Centre, Middlesex University, Enfield, UK

C. Twigger-Ross
Collingwood Environmental Planning, London, UK

In the last few years it has become common practice to discuss flood hazard management and research in terms of 'improving resilience'. Whilst resilience alludes to the idea that floods cannot be prevented and that management must therefore be focussed more towards human adjustment and adaptation, it is nevertheless a highly contested concept. This paper deals with the different interpretations and understandings of resilience which co-exist within the flood hazard management and research communities and explores some of the implications for the future orientation of public policy.

A small-scale study of human response and resilience in post-flood communities was conducted as part of the Environment Agency's 'Improving Institutional and Social Responses to Flooding' (IISRF) research project. The study focused on three communities within Carlisle, which experienced severe flooding in January 2005. Using a mix of data sources, including focus groups, interviews and analysis of published reports and other printed material, the study examined the experiences of people within these communities between January 2005 and November 2007. The aims of the study were to improve understanding of resilience in post-flood communities and to develop recommendations to strengthen the responses of relevant organizations before, during and after major flood events.

The findings from the study showed that residents made their own personal assessments of the flood risk, regardless of whether they received a warning or not. Many of the research participants had a pre-determined idea that a flood would be a slow-onset event and did not react quickly enough to prevent avoidable damage. Furthermore, the nature and severity of the flood impacts varied a great deal and many people experienced lasting psychological problems. Several people were critical of the local authority's response after the flood, although in reality the City and County councils had played important roles in establishing a Communities Re-United initiative which was widely praised by residents. Personal recovery was impeded by difficulties with insurance claims, the effects of shock and a lack of co-ordination among the relevant public organizations. There was a strong sense of fatalism among the research participants and very few had taken actions since 2005 in preparation for the possibility of future flooding. Overall, the study showed that many people in Carlisle are equally, if not more, vulnerable to the effects of flooding now compared to 2005 because emphasis continues to be placed on resilience as 'resistance and restoration' rather than recovery and adaptation. The paper concludes by outlining a number of policy recommendations to reduce human vulnerability and promote more pro-active adjustment and adaptation to flood risk.

Keywords: Resilience, institutional and social responses, Carlisle

Flood, vulnerability and resilience: A real-time study of local recovery following the floods of June 2007 in Hull

R. Sims, W. Medd, E. Kashefi, M. Mort, N. Watson & G. Walker
Lancaster University, UK

C. Twigger-Ross
Collingwood Environmental Planning, London, UK

This paper will present some initial findings from this Environment Agency, ESRC/EPSRC funded project which aims to undertake a real-time longitudinal diary-based study documenting and understanding the everyday experiences of individuals following the floods of June 2007, focusing on the 'forgotten city' of Hull. Focusing on the experiences of local communities in Hull the methodology involves a standing citizen panel of 55 people who are engaged in diary keeping, in-depth interviews, and group discussions.

The presentation will draw on accounts from interviews and diaries to indicate the very different processes people go through in recovering from flood, and highlighting the role of institutional support and investment in the built environment within that. The paper will draw out some of the implications for thinking about resilience, exploring how different manifestations of resilience (as resistance, response, adaptation), are played out in interaction with each other, across scale (household, community, network, zone, city, region) and across social and physical worlds (e.g. community, institution, built environment, critical infrastructure). We will conclude with reflections on the extent to which the recovery process so far entails the development of new forms of resilience and what the implications for developing local level resilience for flood recovery in the future might be.

Keywords: Resilience, vulnerability, qualitative research, diary study

Increasing resilience to storm surge flooding: Risks, social networks and local champions

H. Deeming
Geography Department, Lancaster University, Lancaster, UK

The Pitt Review into the summer floods of 2007 has highlighted the importance of incorporating at-risk populations and communities into local arrangements for flood preparedness and response. However, public engagement with formal flood warning and response systems has until now been very limited, especially in areas exposed to low probability but high consequence flood events (Shaw et al, 2005).

Research into the social and cognitive aspects of flood warning and response indicates that exposed populations are representative of a broad constituency of not only individuals but also formal and informal social networks (Handmer and Parker, 1998). These networks can operate as information conduits but they can also act to filter and adjust hazard and risk information in ways that can either constrain or increase the potential effectiveness of network members' responses from that originally intended by the communication's source.

This paper reports on research in three coastal towns exposed to a low probability sea flood hazard. A questionnaire ($n = 343$) and focus group discussions were used to investigate the role that informal social networks within these towns take in building flood resilience amongst their members.

Whilst the research revealed that discussions amongst informal social networks have recently increased perceptions of risk, it also showed that these perceptions are ameliorated by a very high confidence that the currently employed structural sea defence measures will protect against all contingencies. Broadly, this results in the actions of those who could be termed 'local champions' (i.e. those who act to reduce their own vulnerability to flooding) being ignored. There is also an understanding that during a flood event primary assistance will be forthcoming to all who need it from official rather than unofficial sources; this is contrary to the literature. These findings indicate that more needs to be done to encourage the collective understanding that neighbourhood cooperation is a vital factor in building resilience to extreme events. Suggestions are made as to how such resilience building could be undertaken.

Keywords: Informal social networks, resilience, sea flooding, social capital, extreme event

A new model to estimate risk to life for European flood events

S.M. Tapsell, S.J. Priest, T. Wilson, C. Viavattene & E.C. Penning-Rowsell
Flood Hazard Research Centre, Middlesex University, London, UK

Recent research carried out as part of the EC Floodsite project has focused on the development of innovative models and tools for the evaluation and assessment of the consequences of flooding. With reference to the Source-Pathway-Receptor risk approach to flood risk management, the research discussed in this paper has focused very much on the *receptors* and the risk to human life. The research therefore focused on developing a methodology to estimate loss of life from flood events. In order to reduce the risk to life it is necessary to better understand the causes of loss of life in floods in order to pinpoint where, when and how loss of life is more likely to occur and what kind of intervention and flood risk management measures may be effective in eliminating or reducing serious injuries and fatalities. The objectives of this research were thus:

- to further develop a model, or models, to provide insight into, and estimates of, the potential loss of life in floods, based on work already undertaken in the UK and new data collected on flood events in Continental Europe;
- to map, through the use of GIS and building partly on existing work, the outputs of the risk to life model(s) providing estimates of the potential loss of life in floods.

The research took as a starting point the *Risk to People* model developed in the UK (HR Wallingford, 2003; 2005) and assessed the applicability of this model for flood events in Continental Europe, which tend to be more severe and life threatening. Data on flood events were gathered from 25 locations across six European countries as well as data from an additional case study in the UK. A number of problems were identified with the current model when applied to the flood data collected from Continental Europe. Moreover, research conducted into the factors surrounding European flood fatalities also highlighted the importance of institutional arrangements and mitigating factors such as evacuation and rescue operations, as well as the role of human behaviour.

Thus a new semi-qualitative 'threshold' model which combines hazard and exposure thresholds and mitigating factors has been developed to assess risk to life from flooding in a wider European context. The model has been designed to be flexible enough to be used and applied at a range of scales, from a broad assessment at a regional or national scale, to a more detailed local scale. This flexibility is essential as not all European countries have detailed flood data that is readily available. It is envisaged that the model should be used as a tool to allow flood managers to make general and comparative assessments of risk to life and to consider the targeting of resources before, during and after flooding. The new model also permits simple mapping of risk to life which again can be applied at various scales.

Keywords: Risk to life, injuries and fatalities, receptor impacts, modelling, institutional arrangements, human behaviour

Towards flood risk management with the people at risk: From scientific analysis to practice recommendations (and back)

A. Steinführer & C. Kuhlicke
Helmholtz Centre for Environmental Research—UFZ, Leipzig, Germany

B. De Marchi & A. Scolobig
ISIG Institute of International Sociology, Gorizia, Italy

S. Tapsell & S. Tunstall
Flood Hazard Research Centre Middlesex University, London, UK

The evolving paradigm of flood risk management carries with it a noteworthy reformulation of the idea of public involvement: Beside the well-known demands for (more) stakeholder participation, the at-risk residents themselves are now expected to take active part in reducing the overall flood damage by applying preparatory measures. Hence, a shift of responsibility is obvious which can be referred to as an increasing "individualisation of risk", i.e. a tendency to allocate responsibility for risk reduction and protection also on individuals—rather than to regard it as exclusive to the public sector. These demands for more personal mitigation efforts need to be understood in the light of a general paradigm shift in developed welfare economies worldwide: the one from a top-down oriented government to a governance approach which requires the inclusion of a variety of actors, interests, formal and informal arrangements beyond state-dominated ones.

Yet, based on extensive empirical analyses in Germany, Italy, England and Wales in recent years within the context of the European FLOODsite project (Task 11), in our paper we start from the finding that there is a considerable gap between the scientific understanding of flood risk and its management on the one hand, and the risk constructions of the population, which influence their actions and behaviours, on the other. Our empirical evidence highlights the importance of divergent informal assumptions in risk management. While decision-makers and a number of researchers assume that the public needs and wants to be involved in risk management, large parts of the public are not aware of this demand and regard the allocation of responsibility quite differently. Hence, while participation research and policy all too often start from the tacit assumption that "the people" want to get involved in risk management, be it via preparedness measures, information and awareness campaigns or public hearings, our research findings indicate that the majority of residents neither feel involved nor expect to be involved in decision-making processes concerning flood risk.

In traditional risk management approaches, this 'expert-lay' divide has either been forgotten or taken as a sign of ignorance and disinterest from the part of those at risk—rather than asking, why the latter do not behave in the way they are expected. We therefore regard the ambitious new paradigm of flood risk management as a great challenge for practitioners, policy-makers and researchers since all of them will increasingly be required to talk to and with the people at risk and to listen to them if the actual outcome is to include them in decision-making processes and overall risk governance. We are convinced that this can only be achieved by an integrated approach which leaves the prevailing strict separation between 'objective' risk assessment and 'non-expert' appraisals behind and pays attention to different perspectives and knowledge types.

In FLOODsite, we tried to bridge the above mentioned gap by formulating recommendations for flood risk management with communities at risk—thus translating some of our main findings about risk awareness, actual individual behaviour and social vulnerability into the language of those in charge of flood risk management. This requires the adoption of two perspectives at once: that of the residents at risk and that of the practitioners. A further intention of our paper is also to reflect on this translation process, its impediments and insights, the feedback we received and its implications for future research needs.

Keywords: Flood risk management, risk governance, people at risk

Use of human dimensions factors in the United States and European Union

Susan Durden
US Army Corps of Engineers, Institute for Water Resources, Alexandria, Virginia, USA

C. Mark Dunning
Marstel-Day Consultants, Fredericksburg, Virginia, USA

While "Floods are a social business", as stated by MG Riley of the U.S. Army Corps of Engineers (Corps), flood risk management planning in the U.S. Federal government has given primary focus to enhancing economic well-being as a measure of success. In 2005, new guidance was promulgated in the Corps which requires and emphasizes the inclusion of a broad range of considerations in development of water resources solutions. Arguably the most important of these categories, and, the one most likely to dramatically affect the decision making process, is the Other Social Effects (OSE) account.

The European Union (EU) has a much longer experience of dealing with flooding and its impacts on human populations. There have also been numerous national (e.g. the UK and the Netherlands) dialogues on societal values. As the Corps reinvigorates its consideration of OSE, researchers and practitioners have reached out to their counterparts in the EU. A session is proposed which includes presenters from the U.S. and EU addressing the status, evolution and future of human dimensions as a driving component of flood risk planning and implementation. Generally, the members of the EU have developed explicit policies on flood risk management (FRM) and levels of safety for inhabitants. The U.S. does not have such a framework to inform and guide use of OSE in FRM actions.

This session will reflect the conference theme of "Research into Action" by showcasing research which has yielded tools which are currently in use by those involved in FRM. The individual presentations would be short to allow for discussion among the panelists and with the audience. (Initial interest by participants has been confirmed. The final panel will be established if the session is accepted. If other abstracts on this topic have been submitted, they could be considered for this session.)

Representative panel presentations:

- Composition and Use of the Social Vulnerability Index
  - Dr. Susan Cutter, U. of South Carolina or an associate
- Quality of Life Metrics—A Broad Perspective on What Matters
  - Presenter from the Cadmus Group
- Loss of Life and Social Vulnerability
  - Middlesex University
- Foresight USA

A list of discussion questions around these topics will be developed prior to the session with additional topics solicited from the audience. Perspectives on cultural influences, challenges in consideration of OSE and lessons learned in the U.S. and the EU are some of the topics of particular interest.

Keywords: Human dimensions, other social effects, quality of life, policy, loss of life, frameworks

Double whammy? Are the most at risk the least aware? A study of environmental justice and awareness of flood risk in England and Wales

J.L. Fielding
Department of Sociology, University of Surrey, Guildford, UK

Environmental justice research has come to distinguish between two main approaches to analysis. Much of the original research concentrated on the physical proximity of environmental hazards and disadvantaged sections of society, but more recently attention has also been turned to the origins of such inequalities. The first approach, the association of demographics and environmental hazard, has been described as outcome-based and the second approach, exploring the origins of the associations, as process-based.

This paper explores social inequality within the flood plains of England and Wales in an outcome-based analysis using the UK 2001 Census, surface population models and logistic regression analysis to show significant inequalities exists between the working and middle classes, in all Environment Agency Regions of England and Wales except the Midlands region. This overall inequality is reproduced in both the fluvial and tidal flood plains, although that within the tidal flood plains is especially significant and more pronounced in some areas, especially, in the Eastern regions of England. Whether this inequality is unjust or discriminatory, is a question for debate and research and would necessitate a more process-driven analysis, especially looking at migration into and out of areas in a study of neighbourhood generation. However, while this inequality may not be due to overt discrimination, the social characteristics of the at-risk population are of vital importance, not only in hazard management planning but also as part of any social justice agenda.

Furthermore, these spatial inequalities are contrasted again a secondary analysis of spatially referenced data sets which explored public perception and awareness of flood risk. It is shown that public awareness of risk varies by region. However, while someone may be technically defined as 'at risk', whether they *feel* at risk and are threatened by it will have an impact upon how they respond to that risk. Response to risk was explored by asking 'at risk' respondents what actions they might take when presented with a set of hypothetical flood hazard scenarios.

Not only will knowledge of *who* is at risk inform hazard warning strategies but the perceptions of those at risk and their likely actions will also inform any recovery programme following such a hazardous event.

Keywords: Environmental Inequality, perception flood risk, public response to flood risk

Improving public safety in the United States – from Federal protection to shared flood risk reduction

E.J. Hecker, L.J. Zepp & J.R. Olsen
U.S. Army Corps of Engineers, USA

Historically, the US Army Corps of Engineers (USACE) has played a lead role in managing the nation's flood risks, primarily through engineering solutions to flooding, including floodwater management measures such as levees, reservoirs and dredging projects. Today, USACE is expanding its approach to flood risk management to include a life-cycle, comprehensive and collaborative systems approach that will enable those communities and private citizens that bear the consequences of flooding to participate more fully in sustaining an effective reduction of flood risks. This paper documents the transformation taking place in USACE programs and practices to support this new approach through coordination and collaboration with other agencies as well as improved public education and public involvement.

Responsibility for flood risk management in the United States (US) is shared between multiple Federal, State, and local government agencies with a complex set of programs and authorities. Nationally, both the US Army Corps of Engineers (USACE) and the Department of Homeland Security, Federal Emergency Management Agency (FEMA) have programs to assist states and communities in reducing flood damages and promoting sound flood risk management. However, the authority to determine how land is used in floodplains and to enforce flood-wise requirements is entirely the responsibility of state and local government. Floodplain management choices made by state and local officials impact the effectiveness of federal programs to mitigate flood risk and the performance of federal flood damage reduction infrastructure. Likewise, Federal programs can impact the floodplain use choices made at the non-Federal level of government as well as by private citizens.

For this reason, successful management of the nation's flood risks requires careful and continuous coordination between the Federal and non-Federal levels of government as well as ongoing engagement with the private sector and the public. A key challenge is ensuring that as the public and government leaders make flood risk management decisions, they integrate environmental, social, and economic factors and consider all available tools to improve public safety. This requires a public that is educated both as to the risks they face and actions they can take to reduce their risks.

Historically, the US Army Corps of Engineers has played a lead role in managing the nation's flood risks, primarily through engineering solutions to flooding, including floodwater management measures such as levees, reservoirs and dredging projects. Today, USACE is expanding its approach to a life-cycle, comprehensive and collaborative systems approach that will enable those communities and private citizens that bear the consequences of flooding to participate more fully in sustaining an effective reduction of flood risks.

This paper describes the transformation taking place in USACE programs and practices to support this new approach through coordination and collaboration with other agencies as well as improved public education and public involvement. The first section of this paper provides a backdrop by briefly describing the historical shifts in the Federal and non-Federal roles and responsibilities for flood risk management in the United States. The current transition to a coordinated approach of shared responsibility is then described with examples given of USACE programs that support this transformation.

Evaluating the benefits and limitations of property based flood resistance and resilience – a UK perspective

Neil Thurston & Bill Finlinson
Entec UK, Shrewsbury, UK

Ninoslava Williams & Joscelyne Shaw
Greenstreet Berman, London, UK

John Goudie & Tim Harries
Defra, Flood Management Division, London, UK

At present, there are several hundred thousand properties which are located in significant flood risk areas. Climate change is likely to increase the pressures on these properties and the need for the consideration of a wider portfolio of flood management approaches. This may include the greater use of flood resistance and resilience measures within individual properties.

In July 2007, Entec UK and Greenstreet Berman were commissioned to undertake a research project entitled "FD2607 – Developing the evidence base for flood resilience". This project was designed to provide analytical information relating to the uptake of resistance and resilience measures by households and businesses.

The following paper examines the effectiveness of property based resilience and resistance measures in reducing flood risk over the long term. This paper describes the development of a spreadsheet model which enables the quantification of the costs and benefits of different measures at a property level. This model has been designed to be consistent with established cost/benefit approaches for flood risk assessment (FCDPAG3) and enables the quantification of the benefits of specific measures at different flood depths and return periods.

The paper also considers the barriers to the uptake of property based resistance and resilience techniques. This has been investigated through a stakeholder survey (including insurance, loss adjusters; national flood forum; CIRIA and RICS representatives) and a telephone survey of 1200 residential and commercial properties within significant flood risk areas of England. The paper describes the survey results and examines the significance of different factors (including previous experience of flooding, income etc) in controlling the current uptake of these measures.

The paper concludes by considering policy options for increasing the uptake of property based resistance and resilience measures, including market-based solutions, such as better information and the role of insurance and legislative ones, such as strengthened building regulations.

Keywords: Flood resilience and resistance, Economic cost benefit analysis, Residential and business questionnaire, Policy recommendations

Flood risk management: Experiences from the Scheldt Estuary case study

M. Marchand, K.M. de Bruijn & M.J.P. Mens
Deltares, Delft, The Netherlands

J.H. Slinger & M.E. Cuppen
Delft University of Technology, Delft, The Netherlands

J. Krywkow & A. van der Veen
Twente University, Enschede, The Netherlands

There is an increasing awareness that a valid flood risk assessment requires the involvement of the local public living in the area liable to flooding. Indeed, the new EU Flood Directive stipulates that all stakeholders must be given the opportunity to participate actively in the development and updating of flood risk management plans. Designing and achieving a satisfactory level of public participation, however, remains a challenge. Examples of good practice in participatory flood risk management are still scarce and theoretical guidance is developing slowly. One of the key problems concerns accessing and using different types of knowledge in discussions on flood risks between stakeholders, scientists and policy makers. Moreover, flood risk itself forms only one aspect of regional development plans and policies.

The Scheldt Estuary case study of the FLOODsite project was designed first to explore the role of individual perceptions and knowledge in determining the policy preferences of three actor groupings in the region: the scientists, local citizens and regional and local policy makers, and then to explore how new model-based knowledge influences these preferences. Information was gathered initially through semi-structured interviews with local citizens and scientists, followed by a questionnaire and three different workshops with scientists, policy makers and citizens conducted over a four year period. Model-based scenario analysis was used to generate new scientific insights on future flood risks. Additional insights were garnered from reliability analyses of existing flood defences and evacuation simulation studies.

The case study provided valuable insight in differences and commonalities regarding flood risk perceptions among the different actor groupings participating in the study. We found a body of local knowledge regarding the environment of the Scheldt, including a deep acceptance of living with the risk of flooding amongst citizens. We found a lack of local knowledge amongst scientists and some differences between local citizens and scientists with regard to preferences for future risk reduction measures. There was also a marked difference between respondents from Belgium and the Netherlands regarding their expectations of their governments following a flood.

In explaining the observed differences it is of utmost importance to place the flood risk issue in a broader context. Important contextual and causal factors, which are briefly introduced in this paper, include the international character of the Scheldt Estuary, flowing from Belgium to the Netherlands, the role of EU legislation, such as the Habitat Directive, and the historic flood of 1953. This flood played a decisive role in determining the way flood risk has been was managed in the Netherlands over the past 50 years. Although much of this past policy agenda remains valid, we are able to distinguish that other flood risk management concepts are gaining importance in policy discussions. Signs of transition towards a new policy can thus be identified.

Keywords: Flood risk perception, public participation, local knowledge, flood modelling, transition

Overcoming the barriers to household-level adaptation to flood risk

Tim Harries
Flood Hazard Research Centre, Middlesex University, London, UK
Flood Management Division, Department for Environment Food and Rural Affairs, London, UK

Amongst at-risk households in England and Wales, only 6% of those with no experience of flooding have taken any action to prepare for floods and reduce possible damage, and this figure only rises to 39% for those who do have flood experience Harries (2007). Drawing on PhD research sponsored by the Economic & Social Research Council and the Environment Agency, and on analysis and research performed as part of the policy formulation process at the Flood Management Division, Department for Environment Food and Rural Affairs, this paper explores the reasons for this and discusses some ways in which flood risk adaptation rates can be improved. The paper would complement the one submitted by Entec UK.

The literature contains a number of suggested explanations for people's failure to prepare for events such as floods. Lack of awareness of the risk is one of the more obvious explanations, as is a lack of a complete *mental map* of the risk (Atman et al 1994) and lack of information about possible mitigation measures. Experience of a hazard event is also commonly cited as an important predictor of risk response (e.g. Grothmann and Reusswig 2006). More recently, Harries (2007), has argued that a more fundamental factor is the desire to avoid anxiety and protect a sense of individual and social identity. He asserts that, in situations of flood risk, people prioritize the protection of their immediate *ontological* security over the protection of their longer-term material security. flood risk response is often governed by the desire to avoid anxiety and protect sense of identity and that this often leads to the screening out of information about risk and risk response measures and to a reframing of flood experiences as 'bad luck' and as unlikely to recur.

This paper presents the conclusions of work conducted for Defra to test some of these theories and identify possible ways of overcoming the barriers to household-level flood risk mitigation. Two pieces of primary research are discussed. In the first, a telephone survey by Entec Ltd and Greenstreet Berman Ltd was conducted with 600 at-risk British householders to collect data on their experiences of floods, their behavioural responses to the risk and their feelings about flood risk and the available mitigation measures. In the second, the role of anxiety and social identity barriers was explored in in-depth, face-to-face interviews with householders in areas where public grants had been made available to pay for resistance measures. The policy options considered by Defra for overcoming the various barriers are also discussed, along with some of the considerations that were taken into account in the development of the final policy proposal.

REFERENCES

Atman, C.J., Bostrom, A., Fischoff, B. & Morgan, M.G. (1994). 'Designing risk communications: Completing and correcting mental models of hazardous processes, Part 1'. *Risk Analysis*, 14(5), 779–788.
Grothmann, T. & Reusswig, F. (2006). 'People at risk of flooding: why some residents take precautionary action while others do not'. *Natural Hazards*, 38(1–2).
Harries, T. (2007). *Householder responses to flood risk: The consequences of the search for ontological security* (PhD thesis). Middlesex University: Enfield, Middlesex.

Keywords: Resistance, resilience, public adaptation, policy development

Human vulnerability to flash floods: Addressing physical exposure and behavioural questions

Isabelle Ruin
National Center for Atmospheric Research (NCAR-ISSE). Boulder, Colorado, USA

Jean-Dominique Creutin & Sandrine Anquetin
LTHE (UMR5564, CNRS, IRD, INPG, UJF), Grenoble Cedex 09, France, France

Eve Gruntfest
University of Colorado, Colorado Springs, Colorado, USA

Céline Lutoff
Laboratoire Territoires—UMR PACTE, Grenoble, France

Flash floods surprise people in the midst of their daily activities because they are sudden. They, particularly strike people travelling. For each catastrophe, up to half of the deaths are road users. Hydrometeorological research allows longer prediction lead-times and reduced uncertainty. However, social vulnerability remains an outstanding focus. Experts call for a comprehensive integration of social and natural sciences to improve the understanding of public responses and target loss reduction. A first step in the direction indicated is to better understand the hydrometeorological circumstances of the resulting accidents as well as the behaviour of the population during the crisis.

The 8th and 9th of September 2002 a storm produced more than 600 mm of rainfall in less than 24 hours and triggered a series of flash floods on the Gard River basin in the south of France (Delrieu et al., 2005). This catastrophic event took 23 human lives in 16 distinct sub catchments. Based on this experience, the authors combine analysis of the physical and human response to Mediterranean storms by using both the results of hydrometeorological simulations and qualitative research tools as in-depth interviews of flood victims. Interviews were conducted in the year following the event in order to reconstruct the chronology of people's spatial activities since the beginning of the flood watches to the end of the crisis period.

After details concerning i) the methodology of the behavioural and hydrometeorological analysis, ii) the local context and the event, the authors examine two points: human exposure and behaviours over scales as a critical problem affecting flood risk on the one hand; the dynamics of event distinguishing flash flood and riverine flood response to the storm, on the other hand.

This investigation stresses the specificity of small catchments, which appear to be the more dangerous. It also shows the need for a deeper thinking of post-event investigations and analyses. Usually these analyses further our knowledge within the discipline studied and provide evaluations upon which various types of mitigation and loss reducing practices can be based. Trans-disciplinary contributions are still rare and they tend to be focused temporally, spatially, or institutionally. This contribution to linking social sciences and geophysics constitutes one step in what Morss et al (2005) call the "end to end to end" process that also shows what may be the benefit of expanding those discipline-specific boundaries. We will finally conclude our paper with the presentation of a new project called DELUGE (**D**isasters **E**volving **L**essons **U**sing **G**lobal **E**xperience) that wish to widen existing interdisciplinary and international efforts in substantive and sustainable ways in order to assist practitioners and researchers to reduce losses from short-fuse flood events.

Keywords: Vulnerability, human behaviour, Flash flood, hydrometeorology, un-gauged basins

The first author would like to request the attention of the selection jury for this paper to be considered for the Young FLOODsite prize.

Assessment of extremes

Estimating extremes in a flood risk context. The FLOODsite approach

A. Sanchez-Arcilla & D. Gonzalez-Marco
Laboratori d'Enginyeria Marítima (LIM/UPC), Universitat Politècnica de Catalunya, Barcelona, Spain

P. Prinos
Department of Civil Engineering, Aristotle University of Thessaloniki, Thessaloniki, Greece

The increasing pressures associated to the development of coastal/estuarine zones and those near river courses have enhanced the sensitivity of these areas to natural processes such as flooding or erosion. The characterization of the associated drivers and responses, particularly under extreme conditions, has become an acute need for managing those areas and the risks for uses and resources there located. This paper presents a summary of results obtained in Task 2: *Estimation of Extremes* within the EU Integrated Research Project FLOODsite (contract number: GOCE-CT-2004-505420). The paper deals with the estimation of extremes for risk assessment in fluvial, coastal and estuarine environments. The approach, techniques and conclusions are, however, general enough to be of application to other geophysical variables.

Within the framework of the FLOODsite methodology, and based on the source-pathway-receptor approach, extreme distributions play an essential role in characterizing the sources of risk (Sanchez-Arcilla et al., 2008). These sources can be viewed as boundary conditions which define "loads" on natural or man-made structures. They are, in all cases, associated to the occurrence of extreme events, characterized by a single or a set (joint) of suitable variables. The assessment of risk requires then a balanced estimation of expected damages and a robust quantification of hazards. This characterization, based on the FLOODsite definition of risk (Gouldby and Samuels, 2005) highlights the essential role played by natural extremes in defining risk. The probability of occurrence of such extremes, denoted hazard, is at the core of this paper.

In this work, several statistical models and various fitting techniques are described. Planning an appropriate extremes analysis involves the understanding of the problem to be addressed, selection and preparation of source data, selection of methods for analysis and parameters, and the fitting and use of derived extremes to address the problem. The applications described illustrate some of the pitfalls and difficulties associated with extreme predictions, particularly for the case of more than one variable (Hawkes et al., 2008). Understanding the assumptions and interpreting the obtained results are important parts of the analyst work.

The presented review of statistical concepts and methods will help to illustrate the predictive capabilities of commonly used probabilistic methods as a function of sample size and the inherent variability of the considered variable(s). This will contribute to more reliable hazard assessment and, thus, improved risk-based management and decisions.

Keywords: Coast, Data analysis, Estuary, Extremes, Flood Risk, Probability distribution, River, Statistics

REFERENCES

Gouldby, B. and Samuels, P. (2005), Language of Risk—Project definitions. FLOODsite EU Integrated Project (http://www.floodsite.net), Research Report T32-04-01, 56pp.
Hawkes, P.J., Gonzalez-Marco, D., Sanchez-Arcilla, A. and Prinos, P. (2008), Conclusions and best practice guidance. Journal of Hydraulic Research (in press).
Sanchez-Arcilla, A., Gonzalez-Marco, D., Doorn, N. and Kortenhaus, A. (2008), Extreme values for coastal, estuarine, and riverine environments. Journal of Hydraulic Research (in press).

Inter-site dependence in extremes: Unlocking extra information

Duncan W. Reed
DWRconsult, Cholsey, Oxfordshire, UK

Where long records have been gathered at a relevant site, and are thought to represent a near-stationary process, their statistical analysis is pivotal to environmental risk estimation. In other situations, pooling data from nearby or similar sites provides a pragmatic way of stretching estimates to the long return period typically required. The range of possible estimates is narrowed if the user is guided to a particular distributional assumption (e.g. that annual maximum squared wind speeds follow the Gumbel distribution) and method of fitting (e.g. regionally weighted L-moments).

A well-constructed pooling scheme promotes site-to-site consistency in design values, i.e. estimates of T-year extreme values. This fits well with the "estimate for all sites" mindset favoured by many.

The paper explores the concept that the study of inter-site dependence yields valuable extra information. In this context, dependence is the tendency for extremes at neighbouring sites to occur together. Rank correlation in annual maxima provides a simple measure of pairwise dependence in extreme values at different sites. Where available, event dates can be used to confirm and interpret the patterns of dependence found.

Results are presented from several studies of dependence, including flood data for catchments in the Upper Thames at Oxford and flood data for urbanised catchments in the London area. Sites are mapped by multi-dimensional scaling so that inter-site distance reflects the degree of independence in their extremes. The paper concludes that dependence analysis can inform a wide range of applications.

Keywords: Inter-site dependence, flood risk estimation

The Flood Estimation Handbook and UK practice: Past, present and future

E.J. Stewart, T.R. Kjeldsen, D.A. Jones & D.G. Morris
CEH Wallingford, Wallingford, Oxfordshire, UK

The Flood Estimation Handbook (FEH) was published in 1999 and has effectively become the national standard for flood frequency estimation in the UK. The FEH, which encompasses a suite of methods for statistical frequency analysis of rainfall and floods, was innovative both in its use of digital catchment information and in the way that data from a number of sites can be pooled together to improve the quality of estimates at gauged and ungauged sites. The major benefits of the FEH have been seen in improved risk management and policy implementation, and have had a considerable impact on the UK economy.

This paper considers each of the three key components of the FEH, the rainfall frequency model, the statistical approach to flood frequency estimation and the rainfall-runoff method, and describes how recent and ongoing research is leading to methodological improvements. In the case of rainfall frequency, research related to reservoir safety assessment funded by Defra is currently revising the FEH depth-duration-frequency model for return periods of 100 to over 10,000 years. An Environment Agency-funded project to update the FEH statistical flood frequency procedures has built on theoretical developments together with the availability of flood peak data from HiFlows-UK, resulting in substantial improvements to the predictive capability of the method. Finally, the rainfall-runoff method, which was originally developed as part of the 1975 Flood Studies Report, has been replaced by the ReFH model for design hydrograph analysis.

Other related aspects of the FEH are also being explored, for example the development of new flood risk maps based on the revised methods together with high resolution digital terrain models. Current research is focusing on comparisons with the results of so-called next generation methods based on a continuous simulation approach.

An important aspect of the development of FEH methods is putting the theory into practice and the paper highlights how this is being managed through the development of new FEH software products, as well as allowing the algorithms to be implemented within other widely used packages.

While some key improvements have been made to the existing methods, a number of challenges remain, particularly to improve flood estimation in urban areas and in permeable catchments. Moreover, it is becoming increasingly urgent to consider the likely impacts of climate change on flood and rainfall frequency in the UK and to develop new methods for frequency estimation in a non-stationary environment.

Keywords: Flood estimation, statistical frequency analysis, risk management, digital datasets, ungauged sites

Extreme precipitation mapping for flood risk assessment in ungauged basins of the upper Hron River basin in Slovakia

S. Kohnová, J. Szolgay, K. Hlavčová & L. Gaál
Department of Land and Water Resources Management, Slovak University of Technology, Bratislava, Slovakia

J. Parajka
Institute of Hydraulics, Hydrology and Water Resources Management Vienna University of Technology, Vienna, Austria

The aim of this study was to test and compare methods and tools for spatial interpretation of the maximum daily precipitation totals and their design values. Four approaches to the preprocessing of annual maximum 24-hour precipitation data were used and compared. The methods were tested on the precipitation measurements at 23 climate stations from 1961–2000 in the upper Hron River basin in central Slovakia.

In first approach the direct mapping of at-site estimates of distribution function quantiles was applied. In the second method, the daily measurements of the precipitation totals were interpolated into a regular grid network, and then the time series of the annual maximum daily precipitation totals in each grid point of the selected region were statistically analysed. In the third method, the spatial distribution of the design precipitation was modeled by quantiles predicted by regional precipitation frequency analysis using the Hosking and Wallis procedure based on L-moments and the Region of Influence method. Homogeneity of the region of interest was tested, and the index value (the mean annual maximum daily precipitation) was mapped using spatial interpolation. Quantiles were derived through the dimensionless regional frequency distribution. The kriging interpolation method was applied in all of these three approaches. In the forth method, the estimation of 100-year maximum daily precipitation at 557 stations in Slovakia was the base for the construction of expert's hand drawn isohyets of design maximum daily precipitation totals.

Finally the design values estimated by these methods were compared and the application in engineering practice was discussed.

Keywords: 100-year maximum daily precipitation totals, design precipitation mapping, regional frequency analysis, L-moments, expert's hand drawn map

… *Flood Risk Management: Research and Practice – Samuels et al. (eds)*
© *2009 Taylor & Francis Group, London, ISBN 978-0-415-48507-4*

River flood frequency approaches for ungauged sites

A. Calver & E.J. Stewart
CEH Wallingford, Wallingford, Oxfordshire, UK

An important part of strategic flood risk management is the quantification of probabilities of occurrence of river floods of particular magnitudes for any site, whether or not observations are available from that location. This represents a significant hydrological research challenge in transferring knowledge from areas with data to wider domains where data are sparse.

This paper presents comparisons of results from flood frequency methods established in water industry practice with newly-developing potential next-generation methods. It relates to the UK where current practice largely uses the 1999 Flood Estimation Handbook methods, together with later supplements, and where recent research initiatives for nationwide systems have involved continuous catchment modelling of flood generation responses.

The Flood Estimation Handbook suite of approaches embraces both statistical flood peak analysis methods and unit hydrograph flood event methods. The national continuous simulation research has explored parameter-sparse hydrological modelling and a range of regression and site-similarity approaches to the spatial transference of information.

For a range of over 100 UK catchments, treated as ungauged, the quantitative results from these methods are compared, the (withheld) catchment data being used as a test of the methods (rather than being used in deriving the results).

Keywords: River flood frequency, ungauged sites, Flood Estimation Handbook, statistics of extremes, continuous simulation, catchment modelling

Non-stationary point process models for extreme storm surges

P. Galiatsatou & P. Prinos
Aristotle University of Thessaloniki, Thessaloniki, Greece

In the context of environmental processes, non-stationarity is often apparent because of seasonal effects, perhaps due to different climate patterns in different months, or in the form of trends, possibly due to long-term tendencies. There have been recent attempts to incorporate non-stationarity in the modeling of extreme values. In the present paper the extreme value behavior of storm surges at station Eierlandse Gat (ELd), 20 km off the Dutch coast, is derived from the theory of point processes.

The point process model is based on a characterisation for the pattern of exceedances of a high threshold and has the advantage that the natural parameterisation of the model is in terms of the GEV (Generalised Extreme Value) parameters of the corresponding annual maximum distribution. Working with the point process model means that the issue of stationarity becomes more critical compared to the case of using a model for annual extremes. The point process model provides an opportunity to model seasonality and to incorporate its effects into the extremal analysis, enabling both aggregated and time-specific estimates of extreme behavior. If there is seasonality and it is ignored, inferences based on the point process model are likely to be biased.

In the present paper a time-varying threshold $u(t)$ that has a 1 year period and assures a uniform crossing rate throughout the year is adopted. The threshold is of the form $u(t; a,b) = a + b \cos(2\pi t)$, where t measures the time (years) and a and b are constants that can be interpreted respectively as the median threshold level and the scale of variation between different periods in a year. This choice seems reasonable because of strong seasonal effects in the storm surge data, which generate more exceedances in the winter than in the summer period. To estimate $u(t)$ the year is divided in a winter and a summer period (October–March and April–September) and a fixed uniform crossing rate over these periods is specified.

The point process model in conjunction with likelihood based methods, including Bayesian inference, offer the opportunity to include aspects such as seasonality in the model, to assess the importance of such components and to calculate corresponding risk measures. Except from the time-varying threshold, to account for the annual cycle obvious in the data, the model parameters of location, μ and scale, σ are assumed to vary periodically in time, with a period of 1 year.

A time homogeneous is applied to the storm surge data and its results are compared with those of various models with periodic effects for the location parameter, μ and both the location and scale parameters, μ and σ, corresponding to a cycle-period of one year. Log-likelihood values of different models provide a simple measure of comparison, relative to the number of parameters in each model and reveal evidence of periodic effects. The adaptability of different models to the data is judged using residual probability and quantile plots. Return levels for different return periods are estimated using all different models and compared with each other.

Keywords: non-stationarity, seasonality, point process, time-varying threshold, 1-year period, uniform crossing rate, periodical effects in model parameters, likelihood based methods, return level estimation

Bayesian non-parametric quantile regression using splines for modelling wave heights

Paul Thompson & Dominic Reeve
C-CoDE, School of Engineering, University of Plymouth, Devon, UK

Julian Stander, Yuzhi Cai & Rana Moyeed
School of Mathematics & Statistics, University of Plymouth, Devon, UK

1 ENGINEERING PROBLEM

The successful design of any coastal defence structure which is both reliable and effective can be associated primarily to knowledge of the future conditions which the defence must withstand. This research aims to improve the techniques for producing forecasts of future sea conditions, specifically extreme values. The practical advantage of the procedure is that it allows identification of the wave directions associated with the highest risk to the structure and better estimation of the return levels for wave height.

2 BAYESIAN QUANTILE REGRESSION

Quantile regression (QR) can be utilised to model extreme values as an alternative to classic extreme value theory (EVT) techniques. Quantile regression fits models through a range of quantiles (see Figure 1); hence the entire spectrum of data is modelled easily. In extreme value theory the extremes are modelled separately to non-extreme data meaning separate models are needed. Furthermore, EVT requires the correct specification of extremes before applying EVT modelling techniques which can be incredibly problematic incurring relatively high uncertainty. QR does not require specification of the extremes to model correctly, hence uncertainty is reduced in the model fit, highlighted by the narrow confidence intervals. In this paper we introduce this technique when applied to a coastal engineering application and present the following extensions to this technique. An example result is shown in Figure 2 shows a 90% quantile regression for wave height against wave direction. We can also relate the particular QR curves to return levels; therefore we can create a return level plot based on

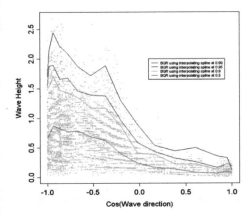

Figure 1.

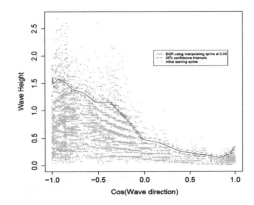

Figure 2.

QR. A further extension to this was to create a 3d return level plot incorporating a direction covariate showing a surface rather than a 2d curve.

One method of performing the QR is Bayesian quantile regression (BQR) techniques developed in Yu & Moyeed (1994) which primarily use a parametric approach to the BQR modelling which provides some excellent results. However, we develop a non-parametric alternative to Yu & Moyeed (1994) using an interpolating spline. The benefit of this approach is the versatility of the non-parametric approach over the parametric when implementing the BQR technique to a range of data sets. The framework for the BQR using a spline is based on altering the Monte Carlo framework used to update points along the interpolating spline rather than parameters from a polynomial model. The benefits of this technique over EVT include reduced computation time, not dependent on correct segregation of extremes, a simple return level plot which can include directional covariate.

Keywords: Bayesian, Quantile Regression, Interpolating Splines, Extreme value theory, Return level plot

REFERENCE

Yu, K and Moyeed, R (2001), "Bayesian Quantile Regression", Statistics and probability Letters, volume 54, page 437–447.

Multiscale probabilistic risk assessment

C. Keef, R. Lamb & P. Dunning
JBA Consulting, Skipton, N.Yorks, UK

J.A. Tawn
Mathematics and Statistics, Lancaster University, Lancaster, UK

In this paper, we discuss methods that can help in answering the questions "How can we calculate the probability distribution of damage from flooding for an entire country?" and "How likely is a certain combination of flooding problems within a given area?"

Understanding and modelling the probability of spatially extensive events is of great importance to government agencies and the reinsurance industry. However, knowledge of the factors affecting this probability calculation is limited. For instance the probability of floods in two neighbouring river catchments is not generally known, and the probability of spatially extensive events and the associated severity of these events also has to be understood.

In order to estimate these probabilities we present a method that is capable of modelling dependence at different spatial scales and at differing levels of extremeness. The approach is capable of accounting for variation in the timing of floods and of handling dependence between a large number of variables (or locations). It is based on a theory of conditional dependence that has been applied by the authors to river flows in the UK (and also by others to storm surge levels in the North Sea). We illustrate the method with examples for river flow in the UK, demonstrating the modelling of the joint distribution of flows at various scales from small catchments through to the national scale.

By simulating data from the statistical model, plausible combinations of extreme (or non-extreme) river flows can be generated for any region. We can then use the simulated data to derive estimates of flood depths using efficient automated flood mapping procedures. Relationships between probabilities of defence failure and flood 'load' can be built into this methodology. Combining the simulated flood data with information about properties and economic impacts, we develop an approach to produce estimates of the distribution of flood damage (or other chosen aggregated measures) for the chosen region.

Keywords: Spatial extremal dependence

Improving the understanding of the risk from groundwater flooding in the UK

D.M.J. Macdonald, J.P. Bloomfield, A.G. Hughes, A.M. MacDonald, B. Adams & A.A. McKenzie
British Geological Survey, Wallingford, Oxfordshire UK

Recent UK Government figures suggest 250,000 properties in the country may be at risk from groundwater flooding. The scale of the problem has been recognised by the Department for the Environment, Food and Rural Affairs (Defra) which is currently assessing how best to map groundwater flood risk and how to incorporate groundwater into early flood warning systems. In addition, groundwater flooding has been included in the EU Floods Directive, although not as a compulsory element.

The characteristic feature of groundwater flooding events is the relatively long duration compared with fluvial flooding, and although limited in extent, these events can as a result cause significant social and economic disruption. Groundwater flooding is defined here as the emergence of groundwater at the ground surface away from perennial river channels, or the rising of groundwater into man-made ground under conditions, where the 'normal' ranges of groundwater level and groundwater flow are exceeded. There three broad hydrogeological scenarios in which groundwater flooding can occur in response to periods of extreme rainfall:

- *Groundwater flooding in unconfined aquifers*: long-lasting, often regionally extensive, groundwater flooding caused by the water table in an unconfined aquifer rising above the land surface. The most vulnerable properties to this form of groundwater flooding are those located on the unconfined chalk aquifers of southern England where groundwater levels can fluctuate over several tens of metres.;
- *Groundwater flooding in alluvial sediments overlying non-aquifers*: relatively localised and more short-lived flooding associated with groundwater levels in alluvial aquifers rising to the ground surface. Alluvial groundwater flooding occurs when water moves laterally out through the permeable sides of a river channel into lower lying alluvial deposits;
- *Anomalous spring flow*: either exceptionally large flows from perennial springs or large flows from intermittent or dormant springs, can cause both localised flooding in the vicinity of the springs and localised flooding down gradient.

In addition to extreme rainfall, groundwater flooding can also occur due to groundwater level rebound, in areas where groundwater abstraction has been significantly reduced, or due to leakage of water distribution pipes, sewers and drains.

The British Geological Survey (BGS) is involved in a number of ongoing and recently completed projects that aim to improve the understanding of groundwater flooding processes. As well as describing in general terms the settings where groundwater flooding occurs, this paper outlines results of these groundwater flooding projects. These include: an assessment of the role of groundwater in flooding in the Oxford floodplain, an area underlain by shallow alluvial deposits; characterisation of unsaturated zone processes related to flooding in the Chalk of southern England; and groundwater investigations to help optimise flood alleviation measures in towns on the Moray coast of Scotland. An approach to map groundwater flooding susceptibility in the UK developed by the BGS is also described, as is an ongoing project to develop a suite of integrated models of climate and groundwater flow in the unsaturated zone and the saturated zone, to help assess and manage risk from groundwater flooding on the UK Chalk.

Keywords: Groundwater flooding, UK

Radar observation of storm rainfall for flash-flood forecasting

Guy Delrieu
LTHE UMR 5564 (CNRS, OSUG) Grenoble Cedex 9 France

Alexis Berne
HWM Wageningen Universiteit, Wageningen, The Netherlands

Marco Borga
DLA University of Padova, Agripolis, Italy

Brice Boudevillain, Benoit Chapon, Pierre-Emmanuel Kirstetter & John Nicol
LTHE UMR 5564 (CNRS, OSUG) Grenoble Cedex 9 France

Daniele Norbiato
DLA University of Padova, Agripolis, Italy

Remko Uijlenhoet
HWM Wageningen Universiteit, Wageningen, The Netherlands

Mountains induce a wide range of meteorological phenomena at the mesoscale, including generation and intensification of precipitation. Furthermore, mountainous topography increases the streamflow volumes and accelerates their concentration. These two factors put a strong emphasis on the requirement for real-time estimation of rainfall in order to mitigate flood and flash-flood hazards in such mountainous areas. The quantitative interpretation of the weather radar signal in terms of rainfall is complex since it depends (i) on the rainfall variability at all scales (scales of the raindrops, of the radar resolution volume and of the precipitating system itself), (ii) on the radar detection domain, constrained by the surrounding relief and the vertical development of precipitations, and (iii) on the parameters and operating protocol of the radar system(s) employed. A pronounced relief obviously adds complexity to the radar quantitative precipitation estimation (QPE) problem by reducing the visibility and increasing environmental noise. This presentation aims at summarizing the results obtained on this subject within the Task 15 of the European FLOODsite integrated project.

A number of innovative algorithms developed by LTHE to identify and correct for various error sources in radar QPE will be presented. This includes (i) use of ground clutter for checking the long-term stability of the radar system; (ii) determination of the radar detection domain by means of geometrical simulations using a digitized terrain model of the region of interest (iii) implementation of a ground clutter identification (GCI) technique based on the pulse-to-pulse variability of the reflectivity; (iv) coupled identification of the rain types and the vertical profiles of reflectivity (VPR), (v) estimation of rain rate at ground level using various combinations of the corrected reflectivity data over the vertical and Z-R relationships conditional on rain types. Such developments are backed on the Bollène-2002 Experiment that was designed to adapt the operation and data processing of the French S-band radar systems for hydrometeorological applications in Mediterranean regions. In addition, a stochastic model of range profiles of raindrop size distributions was developed by the University of Wageningen. This model has been used for testing the sensitivity of attenuation correction schemes to the DSD variability. Such a work nicely complements previous work at LTHE and UniPad aimed at correcting the radar signal for attenuation, a very sensitive effect for X- and C-band radar systems operating in heavy rainfall. These developments represent significant contributions to the so-called radar structured algorithm system (radar SAS) of the partners.

The radar SAS was then implemented and evaluated in the Cévennes-Vivarais (F) and Adige (I) pilot sites devoted to the study of intense rainfall and subsequent flash-flood events. In the French pilot site, various processing strategies (ranging from "static" to adaptive in time and adaptive in time and space) were implemented for the Bollène S-band radar and inter-compared with reference to rain gauge data sets using a geostatistical approach. In the Italian pilot site, a number of algorithms from the SAS library were integrated to develop a radar data processing system specifically suited for Alpine conditions and for use with C-band data. The uncertainty associated with the parameterization of the processing system was characterized following a Bayesian approach.

Keywords: weather radar, quantitative precipitation estimation, hydrological extremes, flash floods

Climate change impact on hydrological extremes along rivers in Belgium

O.F. Boukhris & P. Willems
Hydraulics Laboratory, K.U.Leuven, Belgium

Analysis has been made of potential climate change impacts on hydrological extremes along the main rivers of the Scheldt River Basin District in the Flanders region of Belgium. Based on the simulation of climate change scenarios in hydrological models, prediction has been made of the long-term evolution of the hydrological system in the catchments of the studied rivers.

The climate change impact analysis is based on a continuous simulation approach: The hydrological system behaviour is modeled for an observed historical period and for a future change from the control period (1961–1990) to the predicted period (2071–2100) under forcing of a modified (predicted) climate. The climate change impact on hydrological extremes is assessed through the comparison of key variables of the hydrological system for the two periods (e.g., runoff peaks, low runoff values, runoff volumes, overland flow volumes and actual evapotranspiration).

The modelling procedure is supported by highly resolute regional climate models (simulation results obtained from the PRUDENCE climate project), local scale lumped conceptual hydrological models (NAM of DHI), hydrodynamic models (MIKE11 of DHI) and models for topographical information (DEM: Digital Elevation Models).

An appropriate downscaling method has been developed accounting for the variation of intensity and frequency with time scale. In total, 24 simulations predicting the change in rainfall and potential evapotranspiration, based on 10 different regional climate models, were considered. They were statistically processed in order to obtain potential climate change scenarios for Flanders in the form of low, mean and high variation factors for the variables of precipitation and potential evapotranspiration.

The modelling procedure results state that the predicted climate evolution induces a significant reduction of the low flows due to a considerable hydrological regime modification. As for the hourly high flows (the flood risks), the results range from increasing to decreasing depending on the climate change scenario and thus counting for a large uncertainty. Overland flow peaks and volumes follow similar patterns as for the high flows while actual evapotranspiration shows systematic increase as a result of regional warming.

A statistical method has been implemented for the quantification of the modelling uncertainties induced by the created climate change scenarios and by natural variability. It is based on ensemble modelling of the regional climate model simulations and on Monte Carlo simulations to account for the effect of uncertainties in the selected climate change scenarios due to natural variability when comparing the climate model results with historical data.

Keywords: Climate change, Downscaling, Flood risk, Rainfall-runoff modelling

Uncertainties in 1D flood level modeling: Stochastic analysis of upstream discharge and friction parameter influence

N. Goutal, P. Bernardara, E. de Rocquigny & A. Arnaud
EDF R&D, Chatou, France

The design of dykes for the protection of nuclear power plants against extreme floods (1.000 years return period flood and more) requires hydraulic simulations (1D or 2D).

A major difficulty is that some uncertainties naturally affect not only the hydrological analysis giving the actual discharge value to simulate (Q), but also the actual values of Strickler friction coefficient (Ks), which is calibrated with real data and extrapolated. Here these uncertainties are both evaluated permitting full-scale uncertainty propagation through the one-dimensional free surface hydraulic model. The selected inputs of the model (Q and Ks) are described through a statistical distribution obtained via data analysis and model fitting. Monte Carlo generation is used to produce statistical sample of inputs values to impose in 1D model. Finally, the flood level uncertainty observed as simulation result is analysed via statistical analysis. Multiple interpretations of the results may be considered according to the probabilistic uncertainty modelling of the combination of aleatory and epistemic hydrological uncertainties. This full sensitivity analysis allows us to rank the relative importance of hydrological versus hydraulic uncertainties. An application to a real case study is given and discussed.

Keywords: One dimensional hydraulic model, uncertainty, extremes, floods

*Civil contingency, emergency planning,
flood event management*

Reservoir safety in England and Wales – reducing risk, safeguarding people

I.M. Hope
Environment Agency, Exeter, UK

A.K. Hughes
Atkins Ltd, Epsom, UK

The Water Act 2003 established a new role for the Environment Agency, that of the Enforcement Authority for the Reservoirs Act 1975 in England and Wales. The transfer of this regulatory role from 136 Local Authorities has had a significant impact on the regulated community. Further change is heralded with the forthcoming introduction of Reservoir Flood Plans (i.e. Emergency Action Plans), Post-Incident Reporting and a call to improve current regulations.

Currently there is no legal requirement for reservoir owners or users (undertakers) to produce emergency plans, unlike other high hazard industries.

Defra in its role as policy lead, is promoting the production of guidance on technical standards for Reservoir Flood Plans titled "Engineering Guide to Emergency Planning for UK Reservoirs". The production of Reservoir Flood Plans and their use by the emergency planning community through the Local Resilience Forums will produce greatly increased awareness of the extent of potential inundation areas from reservoirs and challenges to all those participating in the management and communication of flood risk.

This paper highlights the issues involved in:—developing the specification for Reservoir Flood Plans; defining the links with emergency planning authorities through local resilience forums; developing the communication strategy to support the project; highlighting the need for guidance for spatial planning authorities. The paper will also consider how the Ulley incident has informed the approach to the specification for Reservoir Flood Plans.

Keywords: Reservoirs, Flood Plans, Emergency Action Plans

A comparison of evacuation models for flood event management – application on the Schelde and Thames Estuaries

M.J.P. Mens & M. van der Vat
Delft Hydraulics/Deltares, Delft, The Netherlands

D. Lumbroso
HR Wallingford, Wallingford, Oxfordshire, UK

In recent years flood event managers have paid increasing attention to reducing the consequences of flooding by preparing evacuation and rescue plans. Local and regional authorities are often obliged to draw-up flood emergency plans. In many cases these define responsibilities, but details on which evacuation route to take and how much evacuation time is available are missing. In recent years a number of evacuation models have been developed which aim to provide this insight. This paper provides an overview of existing and recently developed evacuation models and discusses how they can be used to assist flood event managers. The models have been tested on the Thames Estuary in the UK and the Schelde Estuary in the Netherlands leading to area specific recommendations for evacuation planning.

The type of evacuation model that is appropriate for a particular flood risk area depends on the magnitude of the flood and the type of area (densely populated urban areas opposed to rural areas). Three scales can be distinguished on which evacuations can be modelled:

- Micro   This corresponds to a scale where each individual receptor at risk (e.g. person, vehicle or property) is modelled;
- Meso   Models used at this scale are often based on movements of traffic streams coupled to relatively simple evacuee route selection logic;
- Macro   A macro scale model provides an initial estimate of the evacuation time for a large area. These often rely upon an aggregate representation of traffic as a series of flows, while attempting to match this demand for road space to the capacity of the road system's links and intersections.

The recently developed evacuation models range from micro scale such as the BC Hydro Life Safety Model to macro scale such as the Evacuation Calculator and ESCAPE. As part of the FLOODsite project, new evacuation models have been developed including a non-linear optimisation model and a simple spreadsheet-based approach. Furthermore the existing dynamic traffic model INDY has been applied to evacuation modelling.

A total of six models has been tried on two pilot areas in the United Kingdom (Thames Estuary) and the Netherlands (Schelde Estuary). In the Schelde Estuary a number of evacuation models were applied on the Walcheren and Zuid-Beveland East and West dike ring areas. In the Thames Estuary models were applied to the Thamesmead embayment and Canvey Island.

It was found that most of the existing evacuation models have been developed for use at a macro scale. This can be useful for making an initial estimate of evacuation times on a regional scale. However, local and regional flood event managers need models that allow for more detail in elements at risk. The end user requires other types of information, such as loss of life estimates, potential road congestion and the effects of management response. Micro- and meso-scale models serve this purpose, however, a micro-scale model of individual receptors often requires a considerable effort to set up which is not always desirable for the evacuation planning of large areas.

Keywords: Evacuation modelling, flood event management, Schelde Estuary, Thames Estuary

The paper should be considered for the Young Floodsite Prize.

Hydrodynamic and loss of life modelling for the 1953 Canvey Island flood

Manuela Di Mauro & Darren Lumbroso
HR Wallingford Ltd, Wallingford, Oxfordshire, UK

Canvey Island is an island in the Thames Estuary, covering an area of 18.5 km². It is formed on an extremely flat and low-lying alluvial fan that has an average height of approximately 1 m below mean high water. Canvey Island is protected against flooding by a network of sea defences. In 1953, the island was inundated by the "Great North Sea Flood" that breached the island's flood defences and resulted in the deaths of 58 people and the destruction of several hundred houses. The consequences of the 1953 floods led to the construction of new flood defences. A concrete wall rising approximately 3 m to 4 m above the mean high water mark currently surrounds Canvey Island. It is claimed that this defence will protect the island against a storm surge level with an annual probability of 0.001 until the year 2030.

As part of the FLOODsite project, work was undertaken to set up both a hydrodynamic and a loss of life model of the 1953 flood. The objective of the work was to obtain a better understanding of the 1953 flood event and to analyse the consequences of breaches in the island's flood defences in terms of loss of life and injuries. The hydrodynamic modelling was carried out using the two dimensional model TUFlow and loss of life was modelled using a Life Safety Model (LSM) developed by BC Hydro in Canada.

In order to model the 1953 Canvey Island flood a historical analysis was undertaken in order to construct the situation that existed at that time. Important sources of information included: historical maps of the island; articles from 1953 newspapers; police reports; and the results of physical modelling carried out in 1954. This information was used to assist in assessing the height and location of the 1953 flood defences, to update the digital terrain model, to reconstruct the tidal shape and to assess the incoming flood volume.

The results of the hydrodynamic model indicated that the 1953 flood covered most of Canvey Island. The model showed that the water depth was 3 m to 4 m at the point closest to the breach with a mean depth of between 0.8 m to 1.0 m. The modelled volume of the 1953 flood was estimated to be 13.1 million m³. This compares well with a 1953 flood volume for Canvey Island of 11.7 million m³ that was estimated by the Kent and Essex River Board shortly after the event.

The TUFlow hydrodynamic model was used as an input to the BC Hydro Life Safety Model (LSM). The LSM models the dynamic interaction between receptors (i.e. people, vehicles and buildings) and the flood wave. The LSM allows estimates of loss of life and damage to buildings to be made. This includes the number of vehicles swept away and buildings destroyed. To assess the loss of life, the population and building data were required for 1953. Data from the 1951 census were used to assess the population of the island in 1953. Historical maps were used to assess the distribution of the properties on the island at the time of the flood.

The results of the reconstruction of the 1953 flood event agreed well with the available historical data. The BC Hydro LSM model indicated that approximately 100 fatalities had occurred during the 1953 event. This number is dependent on the "resilience factors" applied to both people and buildings. The actual number of people that died in 1953 was 58. The number of buildings destroyed during event is unclear. However, the anecdotal evidence available seems to be similar to the LSM model results.

Keywords: Evacuation modelling, hydrodynamic modelling, Canvey Island, 1953

The paper should be considered for the Young Floodsite Prize.

Short-range plain flood forecasting and risk management in the Bavarian Danube basin

M. Mueller & M. Tinz
Infoterra GmbH, Friedrichshafen, Germany

A. Assmann
Geomer GmbH, Heidelberg, Germany

P. Krahe & C. Rachimow
Bundesanstalt für Gewässerkunde, Koblenz, Germany

K. Daamen
Bayerisches Landesamt für Umwelt, Munich, Germany

J. Bliefernicht & C. Ebert
Institut für Wasserbau, Stuttgart, Germany

M. Kunz & J.W. Schipper
Institut für Meteorologie und Klimaforschung, Karlsruhe, Germany

G. Meinel & J. Hennersdorf
Leibniz-Institut für ökologische Raumentwicklung, Dresden, Germany

The European initiative "Global Monitoring for Environment and Security" (GMES) represents a concerted effort 1) to bring data and geo-information providers together with users, so they can better understand each other and 2) to make environmental and security-related information available to the people who need it through enhanced or new services. The objective is to gradually develop and validate a number of pilot operational services, based on selected R & D projects extending and strengthening the current actions. One of those supporting projects is the EU-funded FP6 Integrated Project PREVIEW (PREVention, Information and Early Warning) which aims at developing operational geo-information services in support of European civil protection units and local/regional authorities for the management of risks at the European scale. The risk areas covered are flood, fire, storm, earthquake, landslide and industry. All phases of the risk management cycle are addressed in a consistent and harmonised approach, allowing the exchange of information between the different operators and actors involved. The PREVIEW services are built on integrating Earth Observation data, in-situ

Table 1. Sub-services of the "Short-range plain flood forecasting and risk management" service portfolio.

Sub-service	Description
1. High-resolved weather forecast	Spatially high-resolved (1 × 1 km²) quantitative precipitation simulations using an adapted version of the limited area model COSMO-DE
2. Ensemble short-term flood forecast (lead time <3d)	Probabilistic discharge and water level estimates based on a hydrological ensemble modelling approach addressing the uncertainties stemming from meteorological forecasts as well as from observations
3. Land cover service for water management	Provision of specific land cover products for rainfall-runoff-modelling as well as for flood risk mapping and damage assessment
4. Flood risk mapping and damage assessment	Flood risk and asset maps as well as maps of damage potential derived from land use information and the application of damage functions (possible for different flood events or dam failure scenarios)
5. Flood Information System	Provision of all flood-relevant geo-information data for the users in support of risk management and decision making

measurements, ancillary data, new modelling and recent research and technologies results. Demonstration and validation of the services is done under pre-operational conditions. "Short-range plain flood forecasting and risk management" is a PREVIEW sub-project. Main objective is to develop, demonstrate and validate a prototype of an integrated flood risk management service, supporting flood prevention, forecast and alert, flood risk mapping and damage assessment as well as crisis and post-crisis management. The related sub-services are given in Table 1. Emphasis of the project results to be presented is placed on sub-services 3–5. Key aspect is the integration and operational provision of flood relevant information for a better support of decision makers which is realised by the Flood Information System (FIS, www.floodrisk.eu).

The end users involved are the Bavarian Environment Agency, regional and local services in charge of water and flood risk management as well as civil protection units. Area of interest is the catchment of the River Iller in the Bavarian Danube basin (Germany).

Keywords: GMES, PREVIEW, Danube, flood forecast, flood risk management, flood information, FloodServer

Fast access to ASAR imagery for rapid mapping of flood events

R. Cossu, Ph. Bally, O. Colin & E. Schoepfer
ESA-ESRIN, Directorate of Earth Observation Programmes, Frascati, Italy

G. Trianni
SERCO spa, Frascati, Italy

Earth Observation (EO) is becoming a recognized source of information for disaster management, in response to natural and man-made hazards, in Europe and in the rest of the world. EO based crisis mapping services are generally delivered via projects such as GSE RISK EOS and GSE RESPOND alongside with the International Charter Space and Major Disaster, which enable to provide timely access to crisis data from a variety of EO missions.

The all-weather capability of high resolution Synthetic Aperture Radar (SAR) data provides useful input to crisis and damage mapping. This is particularly relevant for flood monitoring and SAR is considered a useful information source for plain flood events, a frequent and important type of hazard both in Europe and the rest of the world. In this context, it is needed to look at how the access and exploitation of ERS-2 SAR and Envisat ASAR can be accelerated using emerging technologies. Such capability would then help provide rush crisis mapping products combining SAR based observations with other EO crisis data. This investigation is based on the analysis of the requirements of users of GSE RESPOND and International Charter Space and Major Disasters, who come from both the humanitarian aid and the disaster management communities.

This paper investigates how flood monitoring services can benefit from grid technology. An application called Fast Access to Imagery for Rapid Exploitation (FAIRE) has recently been developed and integrated into a Grid-based environment, called Grid Processing on Demand (G-POD). This is a Grid-based environment created at ESRIN-ESA and currently used for both research and operational activities. FAIRE exploits the capabilities of G-POD for both facilitating SAR data access and processing, and providing suitable resources to face computational demanding tasks like coregistration, geoterrain correction, and geographic projections. Users can access the application through a dedicated Web Portal thanks to which it is possible to browse for the required products (specifying the geographical area of interest as well as the acquisition time) and to monitor the whole process.

After preliminary tests on a series of ENVISAT data of the Chinese Poyang lake (selected with advice from GSE RESPOND partner SERTIT, who run flood monitoring tests in the framework of the ESA DRAGON initiative), the system was extensively used for a number of real plain flood events that effected different regions of our planet from July 2007 onwards, including Pakistan, United Kingdom, India, Vietnam, and West Africa. In all the considered cases, the system provided fast access to the original data (both crisis and historical data), coregistered time series, geoterrain corrected data, alongside with additional intermediate products for helping the International Charter and GMES RESOND partners in deriving the final flood maps. Results are usually derived a couple of hours after the crisis image is acquired by the satellite. This activity went along with a validation procedure aimed at ensuring the quality of the results. In this context the validation was referred to the accurate co-registration and geolocation of the image with respect to one another and to a global coordinate system. In all the considered cases the average accuracy was found to be in the order of fraction of pixel, which is acceptable for the addressed application.

Based on both the obtained results and discussions with specialists, it is expected that FAIRE and G-POD can provide a significant contribution to develop an enhanced flood monitoring capability. Users can take advantage of the underlying Grid technology that results in both a transparent access to the huge distributed data archive and a significant decrease of the time required by data processing.

Keywords: Earth Observation, flood management, flood mapping, International Charter on Space and Major Disasters, GMES RESPOND

Benefits of 2D modelling approach for urban flood management

E. David, M. Erlich & A. Masson
SOGREAH Consultants, Water Resources and Environment Division, Echirolles, France

It is now widely recognised that flood risk has to be managed in an integrated manner, including flooding from rivers, the sea, sewer systems and sewer flooding. Considerable progress has been made on the principles of integrated flood risk management. However, the technologies to enable fully integrated risk assessment of flooding from all sources and appraisal of strategic portfolios of options are not yet fully developed in particular case of urban areas, which represent the most challenging locations for broad scale flood risk analysis from multiple sources.

In the paper recent examples of applications of TELEMAC 2D System (www.telemacsystem.com) providing pragmatic approach to attribution of flood risk in urban areas using 2D hydrodynamics will be presented. The key information that is very useful at the various stages of flood event management planning at a local level with a good accuracy (for example in terms of spatial planning allowing recognised flow path around individual buildings) brings a significant insight in this respect. In particular focus will be made on the identification of the critical points related to the representation of the complex topography, geometry of buildings, roads, sewers etc. to be taken into account for an appropriate urban flooding case studies so that they can be targeted for improvement.

Local and borough combined urban developments having for objective to reduce the flood depth around the urban areas where economic and social stakes are high will be presented.

The link between 2D approach, economic stakes assessment and flood crisis management plans for urban areas introduced in France in 2004 through so—called Community Safeguard Plan law, will be assessed and appropriate lessons relevant to the existing European regulations including Flood Framework Directive will be drawn.

Keywords: Urban floods, 2D hydrodynamic modelling, Community Safeguard Plan

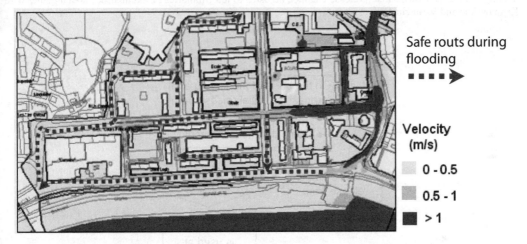

Figure 1. Routes of evacuation during Flooding.

Computer modelling of hydrodynamic conditions on the Lower Kuban under various scenarios and definition of limiting values of releases from the Krasnodar, Shapsugsky and Varnavinsky hydrounits for prevention of flooding

M.A. Volinov, A.L. Buber & M.V. Troshina
The All Russian Research Institute Hydraulics and Land Reclamation named after A.N. Kostiakov, Russia

A.M. Zeiliguer & O.S. Ermolaeva
Moscow State University of Environmental Engineering, Moscow, Russia

Hydrological conditions of Lower Kuban River watershed are extremely variable. Frequently arising floods, alternated with the periods of water shortage. The catastrophic flood events occur promptly and, commonly, there is not enough time to manage flooding by normal operation of reservoirs. The management of water releases using standard rules during high waters is not effective. There is a need to use the imitation hydrodynamic model of the river network for the flood forecasting to check all possible variants of flood development and to take the efficient decision in short time.

The model of the water object, validated and calibrated on the data of observations, allows to determine the basic river parameters (discharges, levels, velocities etc.) in the operative mode and to solve the following tasks:

- Operation of hydroelectric power stations during flood;
- Protection of the floodplain;
- Management of water resources during the period of water deficit.

The object of modeling is the basin of Lower Kuban River, including the following parts: Krasnodarskoe, Shapsugskoe, Krjukovskoe, Varnavinskoe reservoirs, Fedorovskij and Tikhovskij hydroelectric power stations, Lower Kuban River and it's branch Protoka from Krasnodar hydrosystem up to the Azov Sea, Krjukovskij connecting channel and Varnavinskij release channel, adjusted to left tributaries of Zakubanskij area together with Krjukovskoe and Varnavinskoe reservoirs.

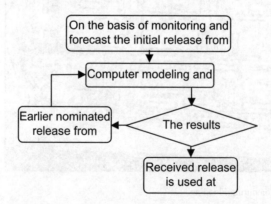

Figure 1. Block scheme of the water system regulation technique.

The multipurpose hydrodynamic model of system of interconnected rivers and channels at Lower Kuban River was developed with MIKE 11 package (Danish Hydraulic Institute). On its basis the special technique of water system regulation to protect agricultural areas in floodplain zone was developed. The general block-scheme is given in Figure 1.

Developed approach along with authentic and regularly peer hour monitored data provide us with necessary tool for qualitative management of regulating hydraulic structures during the passage of high waters.

Keywords: Flood risk assessment, MIKE 11, hydrodynamic model

Flood warning in the UK: Shifting the focus

C.L. Twigger-Ross & A. Fernandez-Bilbao
Collingwood Environmental Planning, London, UK

G.P Walker, H. Deeming, E. Kasheri & N. Watson
Geography Department, Lancaster University, UK

S. Tapsell
Flood Hazard Research Centre, Middlesex University, UK

Flood events over the past five years in the UK have raised the profile of flooding in general and flood warning in particular. Specifically, the floods in Carlisle raised issues about warning older people living alone which put a spotlight on the question of how to warn people with different vulnerabilities. Further, floods in Boscastle (rapid response catchment) brought to the fore the difficulty of warning people when there is little or no time to issue a warning. These cases and others touch on the complex and contextual nature of flooding, making a nationally consistent flood warning system, which is flexible to local needs difficult to achieve.

This paper presents recent research examining research and current practice of flood warning by the Environment Agency in England and Wales. It focuses on the warnings issued by the Environment Agency which are for river and coastal flooding only. The research involved a focussed literature review together with questionnaires, interviews and a workshop with staff from the Environment Agency working directly in flood incident management.

The review of research reinforced the need to tailor flood warning messages, and methods of warning according to type of flood, people characteristics and type of place. This was echoed by findings from the data collected on current practice. Taken together research, practice and events suggest that an effective warning system needs to be flexible to a range of key characteristics, and crucially needs to be focussed on facilitating effective responses in floods rather than focussing on warning large numbers of people. Our examination of the current flood warning system managed by the Environment Agency suggests that its emphasis is on the latter question. The system is designed from a technology perspective because that is what will enable a large number of people to be warned quickly, rather than from a person perspective. Looking at the current system it can be said that it works best for fluvial/tidal floods in predictable catchments where people are motivated to be signed up to Floodline Warnings Direct, have the technology and know how to respond. We suggest that this has happened, partly because of the implicit technoeconomic approach that dominates flood risk management which leads to an emphasis on technological solutions ("pieces of kit"). A technoeconomic approach focuses on the idea that if "technical knowledge is rigorously tested and demonstrably proved, then consumption choices will be made rationally" and is contrasted with "sociotechnical—the idea that science is a sociocultural phenomenon and that the technical is always in relationship with wider social, economic and political processes" (Guy 2004, p. 687). Whilst there is work on social aspects of flood warning there has been an emphasis on forecasting and developing the Floodline Warnings Direct[1] as technical projects rather than as part of social systems.

In this paper we discuss approaches for moving the current flood warning system to a response-centred system focussing on the policy framework, the organisation, and the practical aspects that would need to change.

Keywords: Flood warning, social sciences, England and Wales

New approaches to ex-post evaluation of risk reduction measures: The example of flood proofing in Dresden, Germany

A. Olfert & J. Schanze
Leibniz Institute of Ecological and Regional Development, Dresden, Germany

Flood risk reduction is an essential component of flood risk management. Recent decades have shown the development of various methods for risk reduction which either physically adjust elements of the flood risk system (measures) or which shape the scope for action to reduce the risk (instruments). However, risk management portfolios are still dominated by traditional approaches. New methods often cannot be considered due to a lack of experience with their performance and potential difficulties with implementation.

Analyses of effectiveness and other criteria are most common in the so called ex-ante evaluation, applied prospectively for the selection of options. The retrospective ex-post analysis of the performance of measures and instruments is a general deficit in flood risk management. Analyses after flood events are usually limited to the description of the flooding and losses incurred. The question of how the implemented risk reduction measures have influenced losses in reality during flooding is usually not asked in detail. This leads to a knowledge gap between the implementation of risk reduction and the planning of next risk reduction measures which delays the learning from previous practice. Ex-post evaluation of risk reduction is important to provide the informative basis for better learning in flood risk management.

To enable systematic and comprehensive analysis, a *methodology for ex-post evaluation of measures and instruments* has been developed by Task 12 of the FLOODsite project. It has been tested in four case studies with different types of measures including flood proofing and evacuation of goods, calamity polders, dikes and flood warning.

The paper first presents the developed methodology for ex-post evaluation as a generic framework to be applicable with all measures and instruments. On the one hand, this includes the applied effect indicators and the case specific indicator selection. On the other hand, criteria such as effectiveness, efficiency, robustness and flexibility and methods for their application in ex-post evaluation are introduced.

Second, results from an exemplary ex-post evaluation are presented. The case study focuses on flood proofing and evacuation measures applied by private households and businesses during the April 2006 flooding along the Elbe River in the region of Dresden (Germany). It is shown, that 65% more measures have been applied in comparison to the catastrophic flooding of August 2002. The case study shows in detail which measures resp. portfolios are applied and how they reduce losses. The often high values achieved for effectiveness and efficiency are discussed in the light of context factors such as household income, available information, timeliness of warning, household age and other. It is shown that different groups of flood prone population are more or less able to ensure effective private risk reduction. This finally allows to come up with case specific and policy relevant conclusions about the opportunities and limitations of the evaluated types of risk reduction and the options for public authorities to support private action.

Keywords: Risk reduction, flood proofing, evacuation, ex-post evaluation

Dilemmas in land use planning in flood prone areas

A. Scolobig & B. De Marchi
Mass Emergencies Programme, Institute of International Sociology, Gorizia, Italy

Dilemmas about safety standards abound in floodplain areas. Very often they are rooted in interest and value conflicts, which are not resolvable nor negotiable. This is the case for instance with equity issues in risk distribution and the (supposed) trade-offs between economic growth and safety. Indeed when establishing zoning or developing limitations in floodplain areas, local authorities need to consider individual freedom and private property rights on the one hand, and public safety on the other. Moreover, contrasting symbolic interpretations co-exist of binding rules for land use planning. For example the absence of building limitations in a certain area may be taken as a confirmation of safety from the part of the residents and as a factor of risk amplification from the part of geologists. Similar dissenting views exist with regard to flood protection works. In some cases the latter are perceived as a tangible sign that risk is totally under control, in contrast with expert communications reminding residents of the existence of a residual, irreducible risk.

Very often the residents' optimistic (or trusting) views are framed by the authorities in charge of flood risk management as "misunderstandings" or "wrong perceptions" and attributed to inadequate communication about hydro-geological risk and rules for land use planning in hazardous areas.

The paper will explore these issues on the basis of empirical research in Italy. This was conducted within the FLOODsite project (Task 11) in six locations in the Adige/Sarca and Tagliamento river basins (Northern Italy), either recently affected by flash flood events or at high risk of flooding. In 2005 and 2006 a number of semi-structured interviews and focus-groups were carried out involving "qualified informers" such as officers from provincial services and agencies for civil protection, water resource management, hydrological risk prevention and management. Another part of the fieldwork consisted in surveys via a standardised questionnaire.

With regard to floodplain zoning, survey results show that residents are preoccupied of the likely trade-offs between private selfish interests and collective safety, the former being perceived as consistent with a model of development which privileges intensive land use and uncontrolled urbanization. Many respondents are convinced that developments have been and are allowed in risk areas, because of risk underestimation, economic constraints (cheaper property allotments), actual scarcity of available land or lack of memory of past events. On their turn "qualified respondents" have mixed feelings about how to best handle land use planning instruments and rules. They recognise the need for strict regulations while at the same time showing concern that communication about flood risk (through words or deeds such as risk zoning) may accentuate anxiety, decrease property values and hinder economic development in certain areas, also generating social conflict.

The paper will discuss these findings and their implications also in view of the implementation of the new European Floods Directive[1] according to which Member States shall make available to the public, among others, the flood hazard and risk maps (Art. 10).

Keywords: Safety standards, flood hazard, zoning, developing limitations, equity, risk awareness, risk perception, risk management, land use planning, symbolic communication

Paper wants to participate in the Young FLOODsite Prize.

---

[1] Directive of the European Parliament and of the Council on the assessment and management of floods. For further information see: http://ec.europa.eu/environment/water/flood_risk/index.htm

Emergency management of flood events in Alpine catchments

H. Romang
WSL Institute for Snow and Avalanche Research SLF, Davos, Switzerland

C. Wilhelm
Forest Department Grisons, Chur, Switzerland

Floods and debris flows frequently cause fatalities and significant economic losses in Switzerland. In order to reduce such negative effects, several methods are applied. Land-use management aims to prevent settlement in hazardous areas. Technical and biological measures are widely implemented which strive to reduce the impact and the frequency of hazardous events. Nonetheless, not all risks can be eliminated. For example some technical measures, though possible, may not be economically, ecologically or socially feasible. Therefore, not only preventive measures but also interventions such as temporary deflection dikes are required to limit the damage. To be completely effective, emergency measures have to be in place before an event actually occurs. In determining when best to take the appropriate action and at the same time to minimise false alarms, emergency managers need the best information possible. The crucial point in this regard is time. In the case of small catchments with short reaction times, good emergency management poses a genuine challenge.

Therefore the Canton of Grisons in Switzerland initiated a project called "intervention plan" in 2006. The intervention plan provides relief unit officers and safety managers with the information needed to plan and organize activities and to give priority to particularly vulnerable sites. It was developed mainly to aid fire brigades, which act in the front lines in an emergency event. The plan is based on a flood hazard map, providing thorough information on the hazard situation. It shows times and locations for possible temporary mitigation measures. The plan helps to save time because the intervention can be prepared and practiced before a real emergency situation occurs, and the needed materials such as mobile flood protection tools can be acquired in advance. In addition the intervention plan distinguishes different event phases, suggesting possible further development of the event. Last but not least, an encapsulated form of the plan, printed on both sides of a sheet of A4 paper for each object and each event phase, facilitates a quick and accurate transfer of information.

The intervention plan was developed for several torrents and rivers in the test-site Klosters. In one case, it had already been used in a real event. The expansion to other municipalities of Grisons is planned.

Keywords: Floods, debris flows, emergency management, intervention

Flood forecasting and warning

Flood warning in smaller catchments

H. Romang, F. Dufour, Matthias Gerber & J. Rhyner
WSL Swiss Federal Institute for Snow and Avalanche Research SLF, Davos, Switzerland

M. Zappa, N. Hilker & C. Hegg
WSL Swiss Federal Institute for Forest, Snow and Landscape, Birmensdorf, Switzerland

Floods and debris flows frequently cause fatalities and significant economic losses in Switzerland. In order to reduce such negative effects, several methods are applied. Land-use management aims to prevent settlement in hazardous areas. Technical and biological measures are widely implemented which strive to reduce the impact and the frequency of hazardous events. Nonetheless, not all risks can be eliminated. For example some technical measures, though possible, may not be economically, ecologically or socially feasible. Furthermore, future flood events may increase in severity due to the effects of climate change. Therefore, not only preventive measures but also interventions such as temporary deflection dikes are required to limit the damage. To be completely effective, emergency measures have to be in place before an event actually occurs. In determining when best to take the appropriate action and at the same time to minimise false alarms, emergency managers need the best information possible. The crucial point in this regard is time. In the case of small catchments with short reaction times, good emergency management poses a genuine challenge.

This challenge may be met largely by a site-specific early warning system which provides information about the evolution of possibly hazardous situations. For this reason, a project was initiated three years ago which involved an information and warning system for hydrological hazards in small and medium scale catchments (in the range of 1 to 1000 km^2). The system was founded on significant experience with the avalanche warning system IFKIS in Switzerland, and was therefore named IFKIS-Hydro.

IFKS-Hydro is a system for the collection of data such as weather forecasts, precipitation measurements, discharge simulations and local observations of event-specific phenomenon. In this way, the best possible data is made available which provides for good management of dangerous situations. In addition, IFKIS-Hydro incorporates a web-based information-platform called InfoManager, which serves as a central hub for the submission and oversight of the data. Using this approach is warranted, as in some scenarios timely access to data of the highest quality and in a standard format is crucial.

Special emphasis ought to be given to local information in examining extreme hydrological events. This is gathered by measuring stations in more remote areas on the one hand, and may also involve the work of observers on the other hand. Observers can be flexibly deployed, and even hardly measurable factors such as slope instabilities, floating wood or bed load can be gathered and later interpreted. The inclusion of specialised local personnel is a must as they have so much responsibility, both for the interpretation of the data and the realization of the intervention.

Warning and intervention are important pillars of successful emergency management. Thus there is a keen interest among the stakeholders involved in Switzerland to further advance IFKIS-Hydro and related tools as well as to employ and standardise them in regions prone to floods and debris flows. Among others, the "Gemeinsame Informationsplattform Naturgefahren" GIN (Common Information Platform on Natural Hazards) will be developed in the near future. The platform is to provide simple access to hazard-related data by agencies and research institutes in Switzerland and will enable a quick exchange of information to support effective emergency management.

Keywords: Flood warning, early warning, information system, debris flows

A prototype of road warning system in flood prone area

P.-A. Versini
GRAHI UPC, Barcelona, Spain

E. Gaume & H. Andrieu
LCPC, Nantes, France

The monitoring and management of road networks is a major issue for rescue services during flooding events especially in flash flood prone areas. The roads are both the main evacuation routes from the areas flooded or threatened by flooding and the main access for the rescue means to these areas. Furthermore, roads are particularly unsafe places during flooding especially during flash flooding. 40% of the victims of floods during the last 20 years in the south Mediterranean region of France were car passengers trapped by rising water on roads (Antoine et al., 2001). Likewise, 240 emergency vehicles have been swept away or severely damaged during the 2002 extraordinary flash flood that occurred in the Gard region in France (MEDD, 2004).

As part of the European research project Floodsite, a first prototype for the forecasting of road submersions in flash flood prone areas has been developed. It links a road vulnerability rating method, which takes into account the characteristics of the road and of its upstream watershed, with a distributed rainfall-runoff model, which computes flood discharges produced by the upstream watershed for all the points of the road network exposed to flooding. Vulnerability levels and discharge values are combined to evaluate a submersion risk at each exposed point of the road network and at each computation time step (15 minutes). Three risk levels have been defined to report the results on maps: moderate submersion risk (yellow), confirmed significant risk (orange), almost certain submersion (red).

The Gard region (French Mediterranean area) has been used as the application case study to develop and to test the proposed method. The Gard region is frequently affected by severe flash floods and an inventory of the road points flooded during a forty year period, the PICH, has recently been completed in part of the region, which served as a support for the calibration and the validation of the method.

The approach has shown promising results on five recent events for which maps of observed road submersions were available. The approach appeared able to predict almost 100% of the actually flooded points with a reasonable false alarm ratio—generally less than 20%.

Beyond the specific application of road network event monitoring, these results illustrate that rainfall-runoff models, despite uncertainty in outputs especially on ungauged watersheds, provide very useful information for flood event management on the condition that their outputs are properly processed and reshaped. This opens new perspectives for the development of flash flood forecasting tools.

Antoine J-M., B. Desailly and F. Gazelle, 2001. Les crues meurtrières du Roussillon au Cévennes. Annales de géographie, 622, 597–623.

MEDD, 2003. Crues du Gard, retour d'expérience, Rapport de l'inspection générale de l'environnement, Ministère de l'Ecologie de l'écologie et du développement durable, Paris.

Keywords: Flood forecasting, Flood risk management, road network

Snow and glacier melt – a distributed energy balance model within a flood forecasting system

J. Asztalos & R. Kirnbauer
Institute for Hydraulic and Water Resources Engineering, Vienna University of Technology, Austria

H. Escher-Vetter & L. Braun
Commission for Glaciology of the Bavarian Academy of Sciences and Humanities, Munich, Germany

Snowmelt is an important contribution to Alpine runoff and to the generation of floods in Alpine rivers. Thus, a snow and glacier melt model has been developed as a component of an operational flood forecasting system which is under development for the Inn river in Tyrol.

The model is based on a fully distributed energy balance approach, and internal processes are parameterized. Because radiation energy input is the most important factor for snowmelt in high Alpine regions special attention is paid to the temporal variability of albedo during snow melt. In contrast to many snowmelt models, the decrease of albedo during melt is not modelled as a function of time with the well known aging curve approach of the US Army Corps of Engineers (1956) but as a function of the total energy input the snowpack has received since the last snowfall. This approach seems to meet the physical background of the process better than the aging curve: Alpine skiers know that on the same day snow conditions (e.g. density or albedo) can be totally different on north or south facing slopes, respectively. This difference can be traced back to the different energy consumed by the snow on the two slopes. The point snowmelt model was calibrated and checked against data collected at the research plot of the Commission for Glaciology of the Bavarian Academy of Sciences and Humanities near the Vernagtferner (Ötztal, Tyrol). The model is then applied for fully distributed snowmelt simulations in the headwater reaches of the great southern tributaries of the Inn river.

A distributed model needs to be calibrated and evaluated with distributed data. Distributed measurements of data suitable for verifying such a model type, e.g. continuous measurements of snow water equivalent or of snowpack outflow, usually do not exist. The only distributed information on melt is the existence or non existence of snowcover on the ground. This information can be derived from a series of photographs taken from a definite point in the catchment and rectifying them to a map scale. These depletion patterns can then be compared to the results of the depletion simulations performed by the model. This procedure could be performed making use of photos taken from the Schwarzkögele above the Vernagtferner. They were rectified, and the depletion patterns were identified manually, which was complicated due to the fact that snow and firn had to be distinguished. Corresponding to the observations the model was upgraded to simulate the water balance of the firn as well.

At the present state, the model is driven with parameters calibrated and evaluated with photos of the Vernagtferner and applied to other headwater catchments in the Ötztal region where no photos but runoff measurements of the streams exist. These runoff observations can be seen as an overall performance check of the melt model. On the whole, the model shows reasonable results. For better simulations of the runoff dynamics, however, it was necessary to implement routing algorithms based on cascades of linear reservoirs that allow the routing of meltwater trough the snowpack and along the river reaches to the gauging stations.

The model is now being implemented in the flood forecasting system HOPI (HOchwasserPrognose Inn) developed in the framework of alpS—Centre of Natural Hazard Management in Innsbruck.

Analysis of weather radar and rain gauges for flood forecasting

M.T.J. Bray, D. Han, I. Cluckie & M. Rico-Ramirez
WEMRC, Department of Civil Engineering, University of Bristol, UK

Real time rainfall for flood forecasting is predominately measured by rain gauge network and weather radars. A good understanding of rain gauge network and its integration with weather radars are crucial for real time flood forecasting systems. To achieve this, a study was conducted to analyse rainfall data measured simultaneously by a dense raingauge network and weather radars over an experimental catchment (The Brue catchment) located in the South West of England over a period of six years. The relevant weather variables (wind, temperature, relative humidity at difference elevations) for the major storms are modelled by MM5 (a numerical weather model) using 1km grid simulation. Initially, a combination of Principal Component Analysis (PCA) and Cluster Analysis was carried out to assess the uncertainty and redundancy of the rain gauge network. PCA provides useful information about the state of an existing rain gauge network. If several components are nearly zero, it would indicate there are many redundant gauges in the network (i.e., they are highly correlated with each other) and further savings could be made by using fewer of them. On the other hand, if there are no near-zero components, the network has no redundancy and may not be adequate. This is equivalent to Nyquist Frequency in Sampling theory. This was followed by cluster analysis which was employed to identify the best locations for the given number of rain gauges found by the PCA. Further research was then conducted using the twelve greatest storm events in order to integrate the rainfall measured by the rain gauge network and the weather radar. The weather factors derived from MM5 model were used to aid the integration of rain gauge data and weather radar so that the optimum rainfall information could be obtained by utilising the strengths and avoiding the weaknesses of all the three main information sources relevant to real time rainfall estimations (rain gauge, weather radar and numerical weather model).

Keywords: Rainfall analysis, rain gauge network, weather radar, Principal Component Analysis, Cluster Analysis, Numerical Weather Model

Integration of hydrological information and knowledge management for rapid decision-making within European flood warning centres

Frank Schlaeger, Daniel Witham & Roland Funke
KISTERS Germany, Department of Environmental Informatics Aachen, Germany

Europe has seen a number of significant flooding events over recent years, particularly the flooding over areas of the United Kingdom in the summer of 2007 but also those seen in central Europe during the summer of 2002. In the latter case, a range of recommendations were made for the improvement to the response to serious flooding situations. These recommendations included simplified communication pathways, a rapid and efficient flood warning distribution process and more precise forecast models.

Further conclusions were also drawn including the need to renew the hydro-meteorological network as well as the software required for data acquisition, quality control and analysis. In addition to this there was a requirement for maximisation of information through the setting up of back-up monitoring systems.

The distribution of knowledge during flooding events was also noted as being in need of improvement. This relates to the integration of all available hydro-meteorological data and information into forecasting and warning system. Improvements were also necessary to ensure the rapid and secure forwarding of information to other authorities and services, as well as residents affected and the general public.

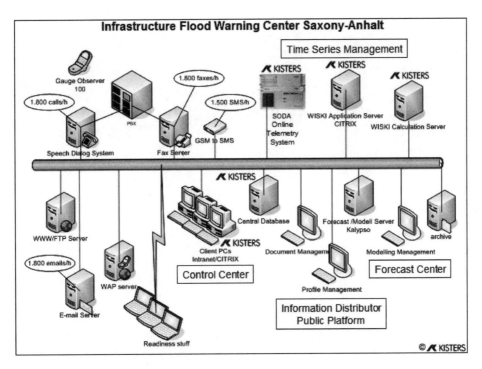

Figure 1. Net scheme and infrastructure of State-Flood Warning Center Saxony-Anhalt, Germany.

This paper will detail the new approaches taken by three European flood warning centres. These are the state flood warning centre in Sachsen-Anhalt, Germany, the flood warning Centre of the Lower Austrian Hydrographic Service, Austria and the DDE Dordogne centre, France. In the case of the flood warning centres mentioned, the KISTERS AG, Germany was contracted to provide all the required components to update the centres.

The restructuring process involved working with governmental organisations and partner companies and successfully achieved the following aims:

1. Optimisation of the telemetry process and communication pathways for real-time collection of hydro-meteorological data.
2. Automation of flood warning alarms and diversification by affected catchments, administrative areas, authorities involved and other interested parties.
3. Consolidation of all types of hydro-meteorological data and information. This includes quantitative data as well qualitative information such as weather forecasts and snow reports.
4. Renewal of flood bulletins and warnings based on available information. Examples include the automatic distribution of messages to predefined groups via varied communication methods such as Email, Fax and SMS.
5. Integration of flood forecasting (models) and other flood information with details of the current situation and potential flood developments.

Details of the technical infrastructure measures in the flood warning centres will be discussed and the results presented. An area of particular interest is the integration of different hydro-meteorological information and forecast results. Additionally, the transfer of this information from the control centre to the publishing of the data to intra- and internet will be covered as well as other methods of data dissemination to specific receiver groups (Fax, SMS, Email, Telephone).

Local warning systems in Slovakia

Danica Lešková, Daniela Kyselová, Peter Rončák & Michaela Hollá
Slovak Hydrometeorological Institute, Bratislava, Slovak Republic

Two pilot local warning systems, based on the information from terrestrial stations, have been established within the POVAPSYS project in Slovakia. Upon the continuous monitoring of changes in precipitation activity development and state of the respective water flow level, the local warning system will transmit warning messages in case of exceeding the advanced setting limits. These messages are intended to a special group of authorized people. The early sending of the warning meassages to competent authorities forms condition for imitiation of securing and rescue works in terms of the Municipal Flood Plan and other related directives, which result from the Flood-Protection Act No. 666/2004 Coll. of Laws of the Slovak Republic.

The terrestrial stations are established after exhaustive terrain research, analysis of previous flood situations, and information from local inhabitants. Localized terrestrial monitoring stations constantly record data from raingauge sensor and water level sensor into a local memory of a monitoring unit and these data are continuously evaluated. In case the previously set limits are exceeded, the local station sends warning message to a central station (CS) and to a personal communication terminal (a cell phone or a pager). Responsible people can analyse the situation through the CS or by direct queries for current values on the monitoring raingauge and water level sensors through the Communication terminal KTO.

The first piloting of LWS Vrbovce was carried out in Western Slovakia in the Upper Myjava River Basin and its confluent, focusing on the GSM data transmission. The second piloting of LWS Čierny Balog, was carried out in the Upper Hron River Basin and its confluent focusing on the radio data communication. Both LWSs are run by municipalities and the SHMU has only a supervisory and advisory function. After testing and expiration of the warranty terms, the LWS will be transposed to the municipalities without payment.

In the LWS Vrbovce, which is a very good working system, we also developed public relations and prepared a flash flood practical exercise with participation of local authorities, the fire department, the river authority, the SMHU, the District and Regional Environmental Offices, i. e. all active flood units, and the non-profit organization of the Global Water Partnership.

The subject of this paper is to present our experience with the operating of the LWS, as well as other activities, like public relations, practical exercises, etc. The aim of piloting the LWS is to meet the needs of those who are interested in carrying out local warning systems (not only municipalities) and the SMHU is ready to offer expert advice and help select localities for monitoring stations, set limit values and solve other problematic situations.

Keywords: Flood warning system, local authorities

The provision of site specific flood warnings using wireless sensor networks

Paul Smith, Keith Beven & Wlodek Tych
Department of Environmental Science, Lancaster University, UK

Danny Hughes, Geoff Coulson & Gordon Blair
Computing Department, Lancaster University, UK

An important aspect of flood risk management is the issuing of timely flood alerts. The spatial, as well as temporal scale, of these warnings is important. In many situations efficient risk management may be aided by the provision of flood predictions at a high spatial resolution. Examples of such situations may include groups of outlying houses or key infrastructure locations such as power sub-stations. In this paper a methodology for providing automated, detailed and location specific warnings which are computed 'on-site' is presented.

The methodology is based upon overlaying and complementing existing observational networks with an adaptive measurement system consisting of wireless sensor nodes. These nodes allow the spatial resolution of the observations to be significantly increased in areas of interest at a relatively low cost. The technology behind the sensor nodes is outlined. Particular emphasis is placed upon highlighting how the adaptive nature of the sensor network can be used increase the usefulness, robustness and manageability of the monitoring system. An example deployment in the River Dee Catchment between Chester and Wrexham is discussed.

Each of the wireless sensor nodes has autonomous computational power. This is utilised to provide local flood warnings. The ability to do this is dependant upon the use of an accurate flood forecasting model which clarifies the uncertainty in its predictions yet has a low computational burden. A suitable family of parsimonious time series models is outlined. The effectiveness of these models is demonstrated using historic data from the River Dee catchment.

Keywords: Flood Risk Management, Flood Forecasting, Uncertainty Estimation, Monitoring Networks

Managing flood risk in Bristol, UK – a fluvial & tidal combined forecasting challenge

M. Dale
Halcrow Group Ltd, Exeter, UK

O. Pollard & K. Tatem
Environment Agency, Exeter, UK

A. Barnes
Halcrow Group Ltd, Exeter, UK

From a flood risk perspective Bristol is a unique city in the UK. Some 5,500 properties are at currently at risk of flooding in the 1% annual probability event from a combination of the Avon and the Frome rivers, from the 3rd highest tidal range in the world which storm surge events can exacerbate and from sewer and surface water flooding. The flood risk picture is further complicated by the presence of the Floating Harbour in central Bristol, a 3.5 km long 'reservoir' connected to the Avon and Frome, and a storm sewer of $72m^2$ area that can convey large volumes of flow away from the city but is dependent on timing of the tide.

The Environment Agency has commissioned four projects in the last three years to manage flood risk in Bristol, including the provision of real time warnings of flooding to the city. This paper reports on the findings of the projects to paint a picture of flood risk in Bristol and to describe how this is being tackled in one of the most complicated urban flood risks locations in the UK. Two of these projects were delivered in 2007 and the other two due for delivery by mid 2008. Project deliverables that are presented and discussed are a real time forecasting model of the entire Bristol Avon catchment incorporated into the Agency's National Flood Forecasting System, a detailed contingency tool to allow flood levels at the Floating Harbour to be predicted if the NFFS is not available, and a 2D overland flow model converted to a 1D model for use with NFFS. The findings and outcomes of these projects are also put into context of overall catchment flood risk reflected in the flood risk management Monitoring and Action plan from the Bristol Avon Catchment Flood Management Plan.

The Bristol case study therefore provides an excellent example of how flood risk is being managed and mitigated in one of the UK's largest cities, in which the tidal, fluvial, pluvial and sewered environments all have a role to play. It presents the flood forecasting challenges which will have resonance with any flood forecasters in cities around the world where combined fluvial and tidal flood risk exists.

Keywords: Flood forecasting, contingency, fluvial-tidal interaction, flood risk management

Off-line flood warning concept for railways

U. Drabek, T. Nester & R. Kirnbauer
Institute for Hydraulic and Water Resources Engineering, TU Vienna, Austria

Many railway lines in alpine regions follow the course of rivers which is often the most economic and reasonable design of railway lines. In the case of flood events, however the railway tracks may be flooded or washed out and there is a possible hazards to passengers as well. In the last years, a series of flood events forced the national railway operator ÖBB to temporarily close down several tracks. The ÖBB decided to commission a project to develop a warning system with an expected lead time in the range of 2 to 4 hours.

The main conceptual formulation of the system is to develop a guideline for setting up an off-line warning system for critical points along a railway line using only precipitation data (forecasts and history) and discharge observations at some gauges along the river reach as a real-time input. This requires the setup of an off-line catalogue consisting of (a) possible catchment response types and future discharge trends at gauges within the next few hours depending on several off-line parameters and (b) information about the expected water level at critical points by routing expected discharges to the points of interest.

The guideline is being developed by setting up the warning system for a sample river reach. The Salzach river from its origin to the first hydro power plant Wallnerau with a catchment area of 2150.5 km^2 has been chosen. Next to the stream there is a main railway line which was partly affected by inundation during past floods. For this reach, a data set of 23 flood events within the years 1996–2005 is available.

The first step in developing the system is to analyse these historical flood events to define characteristic parameters representing flood and catchment response types. The parameters analysed are base flow, duration and slope of the rising limb of the hydrograph, peak discharge, antecedent precipitation index (API), amount of rainfall before and during the event, flood type (rain on snow; snowmelt; convective or frontal precipitation). The result of this step is a matrix of possible catchment responses as a function of parameters that can be easily obtained in real-time. To enlarge the matrix database, a hydrological model (HBV type) has been set up. The antecedent conditions and the input for distinct flood events will be varied to generate a catalogue of typical catchment responses under different conditions at the investigated gauges. Given real-time information about the actual discharge at distinct gauges, recent and predicted rainfall and API, combined with parameters such as base flow at the beginning of the event or season, this catalogue will provide information about the discharge trend expected within the next 2 to 4 hours.

The second step of the system is to route the future discharges at gauges with observations to critical points along the railway line without runoff data (for example bridges, low points which will be inundated first, etc.). Critical points have to be defined together with stakeholders of the railway company and project partners investigating dam and subsoil stability. Off-line calculations of runoff routing will be done via 1-D hydraulic modelling. The result is a catalogue for each critical point. It is planned to estimate a modified discharge rating curve, where discharges of stream gauges and water levels of critical points are combined.

Keywords: Flood warning, off-line catalogue

Satellite observation of storm rainfall for flash-flood forecasting in small and medium-size basins

C. Görner, N. Jatho & C. Bernhofer
Technische Universität Dresden, Institute of Hydrology and Meteorology, Germany

M. Borga
Department of Land and Agroforest Environments (MB), Agripolis, University of Padova, Italy

Flooding—including flash floods—is the most widely distributed of all natural hazards across Europe, causing distress and damage wherever it happens. Previous research has improved understanding of individual factors but many complex interactions need to be addressed for flood risk management in practice.

The "Radar and satellite observation of storm rainfall for flash-flood forecasting in small and medium-size basins" was the topic of Task 15. The aim of this Task was to develop a "Structured Algorithm System" (SAS) for quantitative precipitation estimation (QPE) by radar and satellite at the spatial and time scales of interest for flash-flood analysis and prediction. Task 15 consisted of Action 1: QPE based on radar data (Action leader Grenoble Institute of Technology) and Action 2: QPE based on satellite data (Action leader TU Dresden).

The satellite meteorology team of the TU Dresden took advantage of satellite information to improve rainfall nowcasting in regions of low rain gauge and radar network coverage. TU Dresden developed an algorithm for detecting extreme storm rainfall using the highly resolved geostationary satellites Meteosat-6 and Meteosat-8. Within this algorithm, several techniques for estimating rain rates based on satellite data were applied. The satellite derived rain rates were corrected concerning orographic situation, the wind and moisture conditions and the cloud growth rate by using additional data like MPEF products and radiosondes. Furthermore, they were compared and analysed statistically in relation to the radar data of the co-operation partner of Task 15. The results will be presented for an example of a flash-flood in the Trentino Alto Adige Basin in the Italian Alps which occurred in July 2005.

Keywords: Flood risk management, flash-flood forecasting, satellite observation

Potential warning services for groundwater and pluvial flooding

D. Cobby, R. Falconer, G. Forbes, P. Smyth & N. Widgery
Jacobs Engineering UK Ltd, Reading, Berkshire, UK

G. Astle
Environment Agency, Bristol, UK

J. Dent & B. Golding
Met Office, Exeter, Devon, UK

In response to Defra's Making Space for Water consultation, the RF5 *Feasibility study into expanding flood warning to cover other flood risks* investigated the technical feasibility of providing warning services for sources of flooding other than from rivers and the sea. One of these 'other' sources is pluvial flooding, which can be defined as flooding which results from rainfall generated overland flow *before* the runoff enters any watercourse, drainage system or sewer. As experienced during summer 2007, pluvial flooding can occur almost anywhere and no organisation currently provides a suitably targeted warning service which can assist those managing and responding to pluvial flood risk.

Through initial demonstrations, the RF5 project has determined the technical feasibility of two aspects which would be necessary for any pluvial flood warning service: targeted forecasts of severe storms and identification of topographical areas susceptible to rapid runoff and ponding of extreme rainfall. It is suggested that a form of service could comprise these two main elements:

- Pluvial Flooding Proximity Alert (PFPA): The Met Office could provide information specifically required for management of potential pluvial flooding; i.e. quantitative rainfall forecasts (with associated confidence levels) in a timely manner for a well-defined 'local' area. The proposed PFPA service could fill the gap in the current Met Office forecast and warning services and could be provided directly to the Environment Agency and its partnering organisations. It is envisaged that PEEPs could help to inform surface water management planning.
- Pluvial Extreme Event Plans (PEEPs): PEEPs could be prepared in advance of any event for all urban locations. They consist of five stages which are Screening (to identify the most susceptible locations), Inspection (to verify screening results and identify local risks), Mitigation (improvement of drainage etc.), Contingency Planning (including raising public awareness) and Refinement and Review (to undertake more detailed modelling where appropriate and ensure plans remain up to date). Actions prioritised in the PEEPs could be initiated on receipt of a PFPA to maximise the available lead time.

This trigger rainfall forecast and method for identifying locations most susceptible to pluvial flooding could provide responding organisations with more warning of possible flooding than is currently available.

Keywords: Pluvial flooding, surface water, rainfall forecasting, emergency planning

Data assimilation and adaptive real-time forecasting of water levels in the river Eden catchment, UK

David Leedal, Keith Beven & Peter Young
Lancaster University, Lancaster, UK

Renata Romanowicz
Institute of Geophysics, Warsaw, Poland

The January 2005 flooding of Carlisle (Cumbria, UK) exposed the urgent need for a range of flood risk management measures in this region. The scoping report for the Environment Agency's Eden Catchment Flood Management Plan lists several locations at risk of flooding, as well as providing estimates of the lead time between the causational rainfall process, the issuing of a flood warning, and the eventual inundation event. The report states that for areas protected by a warning mechanism, the warning lead time is limited to a maximum of 2 hours. Any increase in warning lead time could provide a highly useful decrease in human and economic exposure to flood risk. This paper introduces a real time flood forecasting algorithm incorporating data assimilation as a means of extending warning lead times, while maintaining accurate real time level forecasts and providing an indication of forecast uncertainty.

The Eden catchment, Eden and major tributaries are monitored by some 16 tipping bucket rainfall gauges and 31 level (or combined level and flow) gauges. This extensive telemetered monitoring network provides considerable real time information about the state of the catchment. Research carried out by the authors at the Severn catchment site around Shrewsbury (Shropshire, UK) demonstrated that non-linear dynamic transfer function models can be identified from observational data to capture the dynamic relationship between observed rainfall data and river level at key gauging sites along the river. Following the initial model identification and parameter estimation process, the resulting model structure is highly parsimonious and is ideally suited to real time data assimilation and forecasting through a Kalman filter-type mechanism. The complete algorithm is compact, computationally efficient and reliable and easily interfaces with the Environment Agency's Nation Flood Forecasting System. It also provides a means to attach and update confidence intervals to the forecast.

The paper applies the model identification and estimation process and real time data assimilation mechanism to a new data set for the Eden catchment. Importantly, the data includes the 2005 flood event. Additional complexities unique to the Eden catchment and events of the scale encountered in 2005 are addressed. A prototype system is presented for real time forecasting and data assimilation.

Results suggest that the model can generate good quality estimates of level at gauge sites which can be keyed to overbank events. This provides a firm basis for issuing flood warnings. Using this methodology it appears that warning lead times can be usefully extended. Further, the uncertainty methodology inherent in the model identification and estimation and data assimilation process provide a basis for the probabilistic treatment of forecast values. This property is a crucial development given that the discipline of flood risk management is increasingly addressing the flood risk problem within a probabilistic context.

Keywords: Flood forecasting, data assimilation, Eden, Uncertainty

To which extent do rainfall estimation uncertainties limit the accuracy of flash flood forecasts?

L. Moulin
Université P & M. Curie, Paris, France

E. Gaume
LCPC, Nantes, France

Ch. Obled
LTHE, UMR 5564, Grenoble, France

Rainfall-runoff (RR) models are necessary to forecast flash floods. If theoretically, models exists and have sometimes been tested and calibrated in some areas, in practice they are rarely used by operational forecasting services. The main reason for this state of fact is the inaccuracy of the forecasts based on RR simulations that are considered as too high by the hydrologic forecasters used to deterministic and accurate forecasts on large rivers, discrediting the approach in their eyes.

Among the various sources of uncertainties affecting the accuracy of RR simulations, areal rainfall estimation errors seem to play a dominant role. To verify this a priori impression, a thorough analysis of rainfall interpolation errors has been conducted on the upper Loire river catchment (France), an area frequently affected by severe flash floods and covered by a relatively dense rain gauge network. A model of rainfall spatial interpolation errors for hourly rainfall rates based on kriging has been calibrated and validated. This model accounts for temporal dependencies of the errors and is able to produce realistic rainfall interpolation error series.

Based on this model, a large number of rainfall scenarios were produced using Monte Carlo simulations for each of the significant rainfall events that affected the area over the last 15 years. These scenarios were than propagated into a lumped conceptual RR model (GR4), previously calibrated and validated, to assess the impact of the rainfall interpolation errors on the discharge simulations. This procedure has been applied to three catchments with areas ranging from 60 to 3200 km^2, to reveal the possible link between the sensitivity of the model outputs to rainfall estimation errors and the size of the watershed. The realism of the rainfall estimation error model is the main contribution of the presented work if compared to previous studies on the same issue. It makes the strength of the conclusions drawn.

The major conclusion of this study is that RR simulations will probably remain uncertain even in an optimal situation: good quality and long datasets, intensive effort for RR model selection and calibration. A large part of the RR simulation errors can hardly be reduced. It is explained by rainfall estimation uncertainties, except on large watersheds (typically areas over 500 km^2) where the shape of the hydrograph can be influenced by the spatio-temporal pattern of the rainfall event and where distributed RR models may bring a slight improvement if compared to the tested lumped models.

From a practical point of view, rainfall estimation uncertainties limit drastically the possible accuracy of RR simulations. Operational forecasting services should be aware of this limit to efficiently use the RR models and if possible evaluate these uncertainties in real time to be able to deliver confidence intervals along with their traditional deterministic forecasts. Ensemble or Monte Carlo forecasts are now used routinely in meteorological forecasting; there is no reason why they should be disregarded by hydrologists. The error scenario simulation model developed here could help to build such ensemble forecasts in the case where mean areal rainfall amounts are estimated through a rain gauge network. The same type of model is still to be developed for the case where quantitative radar estimations are used.

Keywords: Flood forecasting, Flash floods, rainfall-runoff model, uncertainty

Advances in radar-based flood warning systems. The EHIMI system and the experience in the Besòs flash-flood pilot basin

C. Corral, D. Velasco, D. Forcadell & D. Sempere-Torres
Grup de Recerca Aplicada en Hidrometeorologia (GRAHI), Universitat Politècnica de Catalunya, Barcelona, Spain

E. Velasco
Agència Catalana de l'Aigua (ACA), Barcelona, Spain

The region of Catalunya (North-East Spain) is one of the most affected by floods in the Mediterranean area (generally flash-floods). Owning a dense weather radar network, Catalunya is one of the Hydrometeorological Observatories for the study of flash-floods, and it is involved in several projects related to flood hazard. Inside this region, the Besos catchment (1000 km^2) is one of the selected FLOODsite flash-floods pilot basins.

A big effort is being made by the regional government in order to implement an operational flood warning system, a tool focused on warning decision making for minimising risk to people, economic activity and properties, and guiding water resources management. This has been crystallized in the EHIMI project (Integrated Tool for Hydrometeorological Forecasting), in collaboration with the Catalan Water Agency (ACA, www.gencat.cat/aca) and the Meteorological Service of Catalunya (SMC, www.meteo.cat).

One of the features of this system is its careful radar data processing scheme, which holds several automatic procedures for the tracking between consecutive radar pictures and the correction of signal stability, radar beam occlusion, ground clutter and vertical profile reflectivity. Some of these procedures are zone based, using an identification of areas having different rainfall structures, mainly between stratiform and convective rainfall. The last step of the radar processing is a merging procedure with punctual raingauge measurements from different networks, providing an improved surface rainfall field and several rainfall accumulations related to different periods.

The EHIMI system includes the DiCHiTop distributed rainfall runoff model. It is a grid model based on a loss function at cell scale, provided by a coupled storage model between the SCS model (urban area) and Topmodel (rural and forested). The routing method is based on the linear diffusive wave unit hydrograph, separating the hillslope and the river-channelled process. Currently, a non-linear routing approach derived from the diffusive wave method is being tested in the main rivers.

EHIMI is implemented in the ACA's control centre with successful results. A radar based rainfall nowcasting algorithm has been tested and implemented, with the aim to extend the forecasting lead time with other information (weather forecast models). Inside the FLOODsite framework, the system is oriented to provide distributed warnings based on surface rainfall accumulations and runoff forecasts (at the resolution of 1×1 km^2), having different hydrological sense, and summarising the important places where a risk degree is expected. Currently, these risk warnings are related to hazard probability (synthesised in the return period), but later they will include the concept of vulnerability. Rainfall based warnings are provided in the whole area of Catalunya, while runoff based warnings are only provided where the hydrological model is rightly operational (now in the pilot Besos basin and part of the Llobregat basin), and a methodology to gradually extend the hydrological model to the rest of the region is designed.

Keywords: Flood warning systems, radar hydrology, flash-floods

Flash flood risk management: Advances in hydrological forecasting and warning

Marco Borga
Department of Land and Agroforest Environments, University of Padova, Padova, Italy

Jean-Dominique Creutin
LTHE CNRS, Grenoble, France

Eric Gaume
Laboratoire Central des Ponts et Chaussées, Nantes, France

Mario Martina & Ezio Todini
Department of Earth and Geo-Environmental Sciences, University of Bologna, Bologna, Italy

Jutta Thielen
Institute for Environment and Sustainability, European Commission DG Joint Research Centre, Ispra, Italy

The inherent characteristics of the atmospheric and hydrologic processes involved in flash-flooding provides a severe challenge to the structure of hydrological models conventionally used for flood forecasting and warning. Hydrological modelling of flash flood is made difficult by a number of reasons:

- extreme spatial and temporal rainfall variability, which call for use of weather radar in rainfall quantitative estimation, with the relevant estimation uncertainties;
- uncertainties in rainfall nowcasting, due to the nature and to the limited space and time scale of the relevant atmospheric processes;
- requirement to provide forecasts at river sites for which the hydrologic models have not been previously calibrated, and for extreme and localised events (as flash floods are).

It is therefore unrealistic to expect high levels of forecast reliability for localized thunderstorms occurring on small and medium size watersheds. However, flash flood forecasts do not necessarily need to be accurate to be effective. Indeed, effective flash flood warnings have been issued on the basis of flood warning systems which lack rainfall-runoff forecast models. However, without a rainfall-runoff model it is difficult to forecast the flood potential of storms that have complex space-time texture, particularly when the storm is near the flood/no flood threshold.

The spatially distributed nature of the forecasting problem, on one hand, and the scarcity of data available to calibrate hydrological models (particularly for extreme events), on the other hand, requires examination of the tradeoffs between increased uncertainty and increased spatial specificity when moving to more complex hydrological model structure and smaller modelling elements. A number of different model structures are considered in this study, which summarises the results obtained within Tasks 16 and 23 of the European FLOODsite integrated project. The structures of the forecasting procedures examined here range from threshold-based approaches (i.e. based on use of the approximate threshold basin-average rainfall depth over a given duration that would cause a small stream to begin flooding) to spatially distributed hydrological models. Applications are considered in different hydroclimatic situations, and in particular on Central France, North-eastern Italy and Catalunya (Spain). The study provides an assessment of the effectiveness of using various methodologies for flash flood forecasting and warning and their requirements to enhance preparedness of communities exposed to flash flood risks.

Keywords: Flash flood, forecasting, data quality

Decision support system for flood forecasting in the Guadalquivir river basin

Luis Rein
Confederación Hidrográfica del Guadalquivir

Antonio Linares
BEFESA

Eduardo García & Alfonso Andrés
INCLAM S.A.

The DSS for flood forecasting basin of the Guadalquivir river basin (at the present time, in phase of accomplishment), includes all the scope of its corresponding river watershed (about 58.000 km^2).

It is made up of a set of models—hydrologic and hydraulic—and computer tools that allow users to simulate, in real time, the evolution and behaviour of the water all along the basin. The aim is to feed the system with meteorological and hydrologic inputs obtained automatically by a complex network of sensors (including meteorological radar) distributed trough the river basin that will provide information, at least, every fifteen minutes, as well as with different meteorological forecasts, including ensemble prediction systems (EPS).

The DSS, under development has been conceived as an open system which will be able to elaborate forecasts independently of the hydrologic and hydraulic models used. This will allow the users to adjust the system to the accuracy they need at a given moment and compare the output of the system with those of the different modelling programs used.

The system will use different hydrologic and hydraulic models. From simplified kinematic wave models to bidimensional unsteady models with full Saint Venant equations, including classical one-dimensional models.

Considering the characteristics of the River basin of the Guadalquivir, the DSS also incorporates the operation of existing regulation dams, whose routing capacity has a great importance when mitigating the magnitude of flood peak and volume which would take place in the river basin in their absence.

Keywords: Hydrological and hydraulic forecast, flood management, modelling, real-time

Operational flash flood forecasting chain using hydrological and pluviometric precursors

Giuseppina Brigandì & Giuseppe Tito Aronica
Dipartimento di Ingegneria Civile, Università di Messina, Italy

Flash floods are an important problem in most of the Mediterranean catchments resulting from severe thunderstorm typical of these regions. Real time flash flood forecasting is a key element to improve the civil protection achievements to mitigate damages and safeguard the security of people.

In small catchments, in which flooding occurs rapidly, the forecast lead time may not be long enough to allow the use of rainfall-runoff models that transform past observed rainfall into runoff. So in this case the use of flood precursors implemented off-line and compared in real time with observed or predicted rainfall depths results in a practical alternative.

The procedure here presented is based on the combined use of rainfall thresholds and soil moisture maps; the system is composed of several basic component related to antecedent soil moisture conditions, real-time rainfall monitoring and antecedent rainfall.

Rainfall thresholds are derived using an Instantaneous Unit Hydrograph based lumped rainfall-runoff model with the SCS-CN routine for net rainfall. The chosen has been orientated towards this kind of approach because of its simplicity and particularly because of the reduced number of necessary parameters to calibrate the model.

The soil moisture map can be derived either through antecedent rainfall or through historical flood information and provide the probability to produce an extreme event, given a certain degree of saturation; on the other hand the antecedent soil moisture condition are an indicator for knowing which is the probability to have a flooding with a certain degree of saturation of the soil.

Two different kind of warning can be establish by the system: a first phase of "alert" is identified by a probabilistic soil moisture map derived apart precipitation; a second phase of "alarm" identified by the use of the rainfall thresholds that are a function of the measured or predicted rainfall and of the soil moisture conditions. In a situation of normality, in which it's assumed that the critical limits will not be exceeded, the system is focalized on the analysis of the probabilistic soil moisture map; if the map suggest that we are in a situation of potential risk the alert phase is activated and the use of the rainfall thresholds indicate when the alarm phase have to be activate. Although this procedure could be used for every kind of catchment, it is thought specially for small and medium ungauged basins, where the application of real time model could be difficult to apply because of the lack of data and of the fast response time of the catchments.

A first application of this procedure has been carried out in catchment in Sicily island, Italy

Keywords: Flash flood forecasting, rainfall thresholds, soil moisture

Online updating procedures for flood forecasting with a continuous rainfall-runoff-model

B. Kahl & H.P. Nachtnebel
IWHW, BOKU—University of Natural Resources and Applied Life Sciences, Vienna, Austria

The presented forecasting system consists of a semi-distributed continuous rainfall-runoff model (COSERO) that uses quantitative precipitation and temperature forecasts. Continuous simulation is chosen to provide adequate system states including snow cover and snow water equivalent at the beginning of a forecasting period. This is of particular importance in alpine catchments. I

Rainfall-runoff-models never simulate the observed hydrograph correctly. Deviations of simulated runoff from observed discharge are due to measurement errors, the selected time period for calibration and validation, the parametric uncertainty and the model imprecision. In on-line forecasting systems forecasted input data is used which additionally introduces a major uncertainty in the hydrological forecasting system. To compensate partially these uncertainties different techniques have been elaborated.

The recent study investigates the efficiency of two updating methods in a medium sized catchment in the Austrian part of the Danube basin. The catchment area is about 1000 km². The first method is used for updating the system states at the beginning of the forecasting period. Therefore the precipitation input observed in the last few time intervals is modified by the model to improve the fitting of simulated and observed hydrograph for this time period. The main target is not the corrected precipitation input but the modified state of the system at the beginning of the forecasting period. The rainfall field is changed by a spatially linearly varying correction factor. The parameters of this linear function are obtained by minimizing the differences between simulated and observed runoff for the last few time intervals at the different gauging stations in the catchment.

The second method is an output updating which is directly applied to the forecasted hydrograph. An autoregressive error model is used to update the forecasted hydrograph by analysing the residuals between simulated discharges and incoming observations from the gauging stations for the past few time steps. The AR(1)-Filter eliminates systematic errors and gives better forecasts, especially in the first few hours of the prediction. Additionally the results can be substantially improved by varying the autoregressive parameter with the amplitude of discharge. Between low flow and mean discharges the correlation is much higher than during a flood. Thus the discharges at the different gauging stations are grouped into classes with assigned updating factors. Further improvements of these techniques are achieved by applying the concept in a vector form to all gauging stations by estimating the covariance matrix of residuals at the gauging stations. This provides additionally more stability of the updating procedure, especially when incoming data is partly missing. Both methods proved to be successful and improved the efficiency of the forecasts significantly.

Keywords: Flood Forecasting, Early Warning, Uncertainty

GIS technology in water resources parameter extraction in flood forecasting

V. Ramani Bai & G. Ramadas
School of Civil Engineering, University of Nottingham Malaysia Campus, UK

R. Simons
Environmental Fluids and Coastal Engineering, University College London, UK

Estimation of adequate flood depths in flood modelling is difficult to acquire and so obscure the flood forecasting. New techniques for terrain data collection have been available for the past few years that have potential to provide high resolution data. GIS provides solution to develop data rich environment to hold a database of spatially distributed features and methods to acquire, assemble, incubate, extract, analyse, and view 2D, 3D geographic data. GIS technology is being exclusively devoted to water resources applications and is explored well in this research paper.

This research product is a digital elevation model of Klang basin (Kelang Valley- Figure 1) and the database prepared for the first time in Malaysia with complete perfection and with more major to minor details that is necessary for flood forecasting. It is to retrieve and handle water resources parameters for flood forecasting such as rainfall, type, water level, time of occurrence, ground elevation, hydrologic and meteorological station locations, land use, soil moisture, surface roughness etc (Table 1). This is also a prime model for GIS-based hydrological flood flow using geospatial data more expediently and more accurately than other file handling or manual input methods.

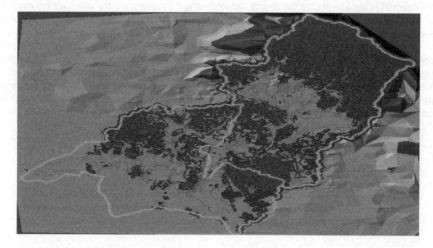

Figure 1. View of DEM of Klang basin.

Table 1. Data used in creation of database of Klang basin for flood modeling.

No.	Data set	Time Period/Scale
	Hydrological Data	
1	Rainfall (Annual, monthly, daily, 10 important events)	1946–47: 2006–07
2	Stream flow (Annual, monthly, daily, hourly 10 correspondig flood events)	1946–47 : 2006–08
3	Water level	1946–47 : 2006–09
4	Storm events	1946–47 : 2006–10
	Digital and spatial data	
5	Topo maps (contours, drainages & other features)	1:25,000
6	Topo maps (Most urbanised: Kuala Lumpur)	1:5,000
7	Land use map	1:50,000
8	Soil map	1:500,000
9	Landsat 7 ETM+ 127/58 Klang valley river basin	Quarterscene cell size 28 m
	Aerial images	
10	Aerial map of Klang river basin	1:20,000
11	Location map of Hydrometry and meteorological stations	1:20,000

Keywords: Flood forecasting, distributed model, digital elevation model, geo-processing spatial data, hydrology, LANDSAT images and flood depth

Combining weather radar and raingauge data for hydrologic applications

Cinzia Mazzetti
ProGeA Srl, Bologna, Italy

Ezio Todini
DSTGA, University of Bologna, Bologna, Italy

In 2001 the paper titled "A Bayesian technique for conditioning RADAR precipitation estimates to raingauge measurements" (*Todini, 2001*) introduced a new technique based upon the use of block-Kriging and of Kalman filtering to combine, optimally in a Bayesian sense, areal precipitation fields estimated from meteorological radar to point measurements of precipitation, such as are provided by a network of rain-gauges. Block Kriging was used to estimate the average field over the radar pixels and its variance from the point rain gauge measurements, while a Kalman filter was taken to find the *a posteriori* estimates by combining the *a priori* estimates provided by the RADAR with the block Kriged measurements provided by the gauges, in a Bayesian framework.

The Bayesian combination technique, was first tested on theoretical synthetic data to assess its convergence to the true known solutions. Successively it was applied to real data and many new requirements had to be faced, such as non-negative weighting and incorporation of errors in gauge measurements, in order to respond to the request of more reliable quantitative and spatial descriptions of the rainfall field.

An original Block Kriging approach to the problem of spatial interpolation was introduced, which also included a new formulation of Kriging with uncertain point precipitation measurements. A Maximum Likelihood estimator was used at each step in time to estimate the semi-variogram parameters, while a new non negativity constraints were added to the Kriging system to prevent negative values in the Kriging weights. The proposed approach showed to be more effective in rainfall estimation than the traditional one, thus preventing the typical Kriging overestimations, smoothing and negative rainfall estimates.

Last but not least, the use of anisotropic variogram functions was incorporated to the Block Kriging formulation in order to better reproduce rainfall patterns.

The paper summarizes the new potentialities of the new system and shows the results obtained in real world applications.

Keywords: Raingauge measurements, weather RADAR, Satellite Precipitation Estimates, Bayesian Combination, Block-Kriging, Kalman Filter, Hydroloigcal Applications

The worst North Sea storm surge for 50 years: Performance of the forecasting system and implications for decision makers

K.J. Horsburgh & J. Williams
Proudman Oceanographic Laboratory, Liverpool, UK

J. Flowerdew, K. Mylne & S. Wortley
Met Office, Exeter, UK

Storm surges are the sea level response to meteorological conditions. On the morning of 9 November 2007, the east coast of the UK experienced the worst storm surge for 50 years. It was initially feared that this event would be as bad as the catastrophic North Sea storm surge of 31 January 1953 that was responsible for 307 fatalities in England and over 1800 in the Netherlands. In the recent event, winds peaked at 38 m/s in some places over the North Sea but they were slightly less than the most extreme forecast. Despite some minor flooding in East Anglia, surge levels from The Wash to The Thames were approximately 20 cm lower than the worst predictions. Storm surges are a natural hazard for which effective forecasting procedures can be used to deliver early warnings and protect lives and property. Warnings of the 9 November 2007 event were in place four days ahead, and accurate, model-based information in the 48 hours leading up to the flooding was delivered to the UK Environment Agency by the Storm Tide Forecasting Service (STFS).

This paper analyses the utility and accuracy of those warnings (the average forecast error was 17 cm). We also provide the hindcast surge (forced by a meteorological reanalysis) as input to a flood inundation model of East Anglia, and we compare the extent of flooding with that observed. During this event, results were also available from a prototype ensemble surge prediction system which is being trialled to demonstrate the benefits of probabilistic forecasting. Throughout the forecasting period, we show that the ensemble spread captured the observed surge at all times. Finally, we couple the most extreme ensemble member to the inundation model and estimate the flood damage that would have been experienced had this scenario been realized. We then perform cost/loss calculations to illustrate the utility of basing flood management decisions (e.g. evacuation, barrier closure) on probabilistic tools.

Keywords: Storm surge, North Sea, inundation, ensemble forecasting

Probabilistic coastal flood forecasting

P.J. Hawkes & N.P. Tozer
HR Wallingford, Wallingford, Oxfordshire, UK

A. Scott
Environment Agency, Peterborough, UK

J. Flowerdew & K. Mylne
Met Office, Exeter, UK

K. Horsburgh
Proudman Oceanographic Laboratory, Liverpool, UK

The Environment Agency has responsibility for fluvial and coastal flood forecasting for England and Wales. The Met Office has operational responsibility for offshore forecasting for the UK. Use of offshore forecasts to estimate the likelihood of high overtopping as an indicator of coastal flooding is not trivial, involving transformation of wave forecasts through the nearshore and surf zones, and the combined effects of wind, waves and sea level in causing overtopping; with sufficient accuracy and reliability for acceptance, and sufficient lead-time for actions to be taken to reduce potential losses.

The project to be described in this paper includes development of surge ensemble modelling, and demonstration of probabilistic coastal flood forecasting for an area in the Irish Sea. Its specific purpose is to develop, demonstrate and evaluate probabilistic methods for surge, nearshore wave, and coastal flood forecasting in England and Wales, but the concepts and models would be equally applicable elsewhere. The main features that distinguish these methods from existing practice are in the use of hydraulic models extending from offshore, through

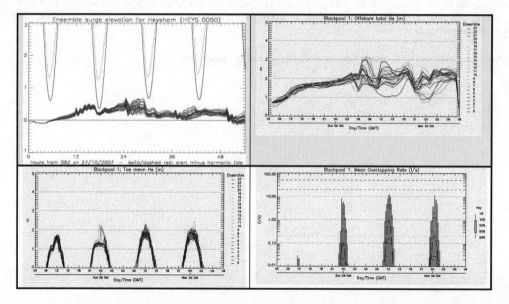

Figure 1. Example surge ensemble and coastal flood forecasts for 27–29 October 2007: a) surge ensemble, b) offshore wave height ensemble, c) toe wave height ensemble, d) probabilistic overtopping rate.

the nearshore and surf zones, to action at coastal defences, and the use of ensemble and other probabilistic approaches throughout.

At the time of writing the FLOOD*risk* 2008 paper, the development and demonstration phases of the project will have finished. This would be the first paper on the project as a whole, although overall evaluation and reporting will not finish until the end of 2008.

Keywords: Coastal, ensemble, waves, surge, overtopping, forecasting

Coastal flood inundation modelling for North Sea lowlands

S. Burg, F. Thorenz & H. Blum
Lower Saxony Water Management, Coastal Defence and Nature Protection Agency, Norden, Germany

RELEVANCE OF LINEAR TOPOGRAPHIC STRUCTURES TO NUMERICAL FLOOD MODELLING AND INUNDATION CHARACTERISTICS IN COASTAL LOWLANDS

Flood risk has been of growing importance in the last years and several aspects of flood risk management have been investigated. However, especially in coastal lowlands research and cooperation between stakeholders and administration is still necessary to meet future challenges. In this context, the project SAFECOAST was built up within the framework of the EU-funded Interreg IIIb North Sea Program. In SAFECOAST 9 partners from the North Sea riparian states are cooperating to deal with the question "How to manage our North Sea coasts in 2050?". Flood risk assessment and flood risk management are covered in different integrated activities, called Actions. Action 5b is carried out by the NLWKN as one of the German partners.

Flood damage evaluations were conducted at two pilot sites in Lower Saxony for the present state and for the state of 2050. For this purpose, a numerical flood model was developed, using the software Sobek. Major flooding was assumed to occur only due to dike breach caused by storm surges, since no big rivers exist in the pilot sites. The actual flood damage evaluation was based on a geographic information system. Two major aspects of the study conducted by the NLWKN were the understanding of flood development and propagation as well as a subsequent assessment of flood mitigation measures for coastal lowlands. In this respect, a parameter study was done and the influence of the specific topography was investigated. Especially the relevance of linear topographic structures was analysed. These are defined as long one-dimensional elements, lying significantly below or above the surrounding surface. Elements below are represented by the drainage or river network; while dikes and street or railway dams compose the major part of the over ground linear elements (see Figure 1).

The effect of linear structures on flood propagation was found to be significant. The drainage network has an accelerating effect and directly transports the flood to low lying parts of the hinterland. On the other hand, the closure of gaps in dams and the alteration of dams are suitable measures to mitigate flood impacts and to influence flow paths. Although important knowledge about flooding of coastal lowlands was obtained from this study, the effects of long linear structures and their influence on flow paths need to be investigated further and for every single project area.

Keywords: Flood inundation model, coastal lowlands, linear structures, Sobek, transnational EU-project

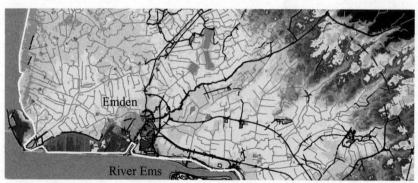

Figure 1. Detail of the study area including main dike (white), dams (black) and drainage network (grey) as linear structures.

New north east of England tidal flood forecasting system

A. Lane
Environment Agency North East Region, Leeds, UK

K. Hu
Royal Haskoning, Haywards Heath, UK

T.S. Hedges
University of Liverpool, Liverpool, UK

M.T. Reis
National Civil Engineering Laboratory, Lisbon, Portugal

People who suffer from flooding are clearly subject to physical danger during the event. Many victims also describe the stress, disruption and unhappiness which they feel long afterwards. Unfortunately, changes in our climate are resulting in rising sea levels and more severe storms, increasing the probability of coastal flooding. However, with sufficient warning of sea defences being overwhelmed by waves, the emergency services can be alerted, local government authorities can take appropriate action, and the general public can prepare themselves. For these and other reasons, the Environment Agency (EA) of England and Wales is investing over £200 million in flood warning systems over the ten year period to 2012–13.

The paper describes the EA's scheme to warn of coastal flooding from Berwick-upon-Tweed to the Humber estuary. Following full-scale trials during winter 2006–07, the system (called TRITON) has now been formally launched by the EA, North East Region, as part of the National Flood Forecasting System (NFFS). TRITON is a nationally adopted coastal forecasting system and this paper describes the system configured specifically for North East Region. North East TRITON went online just in time to warn of the flooding in November 2007 at Immingham, Scarborough and Sandsend, caused by a 3 m tidal surge (as recorded in Thames region) and gale force winds in the North Sea. Before the event, the agency had compared these conditions with storms in the winter of 1953 when a 3 m tidal surge flooded the east coast of England, killing more than 300 people. Luckily, the 2007 surge did not coincide with high tide, reducing enormously the impacts and consequences of flooding.

The new system enables flood experts to more accurately predict, monitor and warn people about tidal flooding, using the latest advances in coastal modelling. The system has been configured to manage climate change and sea level rise. Warnings to householders and businesses can be delivered via telephone, text message, email, fax or pager. The main features of the new system are:

- it transforms offshore wave forecasts to inshore locations, including estuaries;
- it transforms inshore wave forecasts to mean wave overtopping forecasts;
- it adopts dual flood warning criteria or triggers;
- the flood warning maps consider land use and flood risk;
- the flood warning maps and triggers are inter-linked;
- it uses pre-run matrices so that the system is computationally efficient and real-time reliable;
- it uses TRITON and 'Test Harness' data processing software (the former is compatible with the NFFS and it forecasts inshore wave conditions and mean wave overtopping rates by interpolation between the matrices; the latter can be run stand-alone in case the EA flood warning systems fail).

The paper begins by presenting background information. It then explains the adopted methodology and approaches in detail. Finally, it describes the full-scale trials of winter 2006–07 and compares the system predictions to the measurements recorded during recent flooding in the region.

Keywords: Flood risk management, coastal flood forecasting/warning system, north east of England

Impact of extreme waves and water levels in the south Baltic Sea

H. Hanson & M. Larson
Department of Water Resources Engineering, Lund University, Lund, Sweden

In the south of Sweden the landscape is low-laying and the coastline is typically made up of sandy beaches. In addition, a relative sea-level rise makes the coastal areas here generally subject to flooding and erosion (Larson and Hanson 1993). Such events depend on the magnitude and frequency of water levels and wave heights as well as their joint occurrence. The main objective of this study is to jointly analyze waves and water levels, based on existing climate data, in order to assess the risk of extreme events occurring in the southern Baltic Sea with large consequences regarding flooding and erosion.

Long time series of climate data from the south coast of Sweden (3-hourly wind data from Falsterbo 1961-2004 together with daily max, min, and mean water level over the period 1887–1986 from Ystad) were employed to investigate the statistical properties and long-term trends of extreme events in terms of the waves and water level changes. Joint distributions of waves and water levels were obtained from combined time series of wind and water level covering the period 1982 to 2004, with values every three hours, encompassing in total 66,481 values. Through the combined analyses of waves and water levels the probability of extreme events with respect to run-up levels (obtained as a combination of water levels and wave run-up) occurring in the southern Baltic Sea was assessed. The study also established relevant probability distributions to characterize such extreme events as a basis of various risk assessments related to the impact on the coastal areas of large storms.

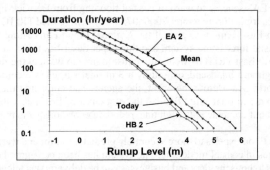

Figure 1. Hours per year of exceedence of run-up levels in Ystad for different scenarios.

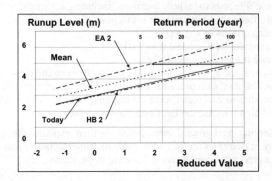

Figure 2. Comparison of return periods of run-up levels for the different climate change scenarios.

Furthermore, an attempt was made to estimate the conditions after climate change. Based on available forecasts and scenarios of future climate change (IPCC EA2 and HB2, respectively, as well as the mean of the two), the corresponding probability distributions were determined. Figure 1 shows present and estimated future hours per year of exceedence of run-up levels simulated for Ystad for different scenarios (levels are referenced to the present mean sea level). As seen from the figure scenario HB2 implies less flooding compared to the present conditions, whereas the mean and EA2 scenarios cause a significant increase in the duration of the flooding for a certain elevation. A comparison with respect to the return period for a certain annual maximum run-up level between the present conditions and the future scenarios is shown in Fig. 2, with fitted Gumbel distributions for the different scenarios as well as for the present conditions. At present, a 100-year return period yields an annual maximum run-up level of about 4.9 m. The corresponding value for the EA2 scenario is 6.2 m. The future return period that corresponds to the present 100-year value is for the EA2 scenario only about 8 years, *i.e.*, flooding at this level will be about thirteen times more frequent in 2100 than what it is today.

Keywords: Baltic sea, Climate change, Extreme waves, Flooding, Statistical analysis, Storm surge

Bayesian rainfall thresholds for flash flood guidance

Mario L.V. Martina & Ezio Todini
Department of Earth and Geo-Environmental Sciences, University of Bologna, Bologna, Italy

The aim of this work is to determine the FFG rainfall depth and compute FFG-values based on the minimization of a Bayesian Loss Function of the discharge conditional upon the state of saturation of the catchment. Rainfall thresholds are here defined as the cumulated volume of rainfall during a storm event which can generate a critical water stage (or discharge) at a specific river section. When the rainfall threshold value is exceeded, the likelihood that the critical river level (or discharge) will be reached is high and consequently it becomes appropriate to issue a flood alert; alternatively, no flood alert is going to be issued when the threshold level is not reached. In other words the rainfall thresholds must incorporate a "convenient" dependence between the cumulated rainfall volume during the storm duration and the possible consequences on the water level or discharge in a river section. The term "convenient" is here used according to the meaning of the decision theory under uncertainty conditions, namely the decision which corresponds to the minimum (or the maximum) expected value of a Bayesian cost utility function. There are described two possible approaches for the same methodology: (a) using the Monte-Carlo simulations or (2) using the Normal Quantile Transform. The main difference of the two is the requirements in terms of data, i.e. the timeseries of rainfall and discharge. Application of the methodology and comparison with other methodologies are provided for the Posina catchment in Italy.

Keywords: Flash flood, rainfall thresholds, Bayesian approach, Flood warning systems

"This paper should be considered for the Young FLOODsite Prize"

Environmental impacts, morphology & sediments

Assessment of hydraulic, economic and ecological impacts of flood polder management – a case study from the Elbe River, Germany

S. Förster & A. Bronstert
Institute of Geo-ecology, University of Potsdam, Golm, Germany

Flood polders are part of the flood risk management strategy at many large lowland rivers. They serve the primary purpose of controlled water storage during large flood events in order to decrease the flood risk downstream along the river. Flood polders are typically dry storage basins that are used for agriculture or other land use of low economic and ecological vulnerability.

The effectiveness of utilising a flood polder in terms of peak attenuation is influenced by several factors such as the storage capacity compared to the discharge volume of the flood wave, the location of the flood polder with respect to the areas to be protected, the operation scheme of the control structures, the quality of the flood forecast and the shape of the flood wave. The cost-effectiveness of the measure is mostly affected by the prevented damage in the downstream areas or reduced expenses for flood management, the costs for construction and maintenance of the flood polder dikes and control structures, damage occurring due to the temporary water storage and the probability of utilisation of the flood polder. Apart from economic damage, utilisation of flood polders can lead to environmentally harmful situations in the storage basin. One major problem that has been observed during long storage times is the depletion of dissolved oxygen concentrations due to the strong oxygen demand imposed by organic material in the water body and at the bottom. This may cause stress on the aquatic ecosystem and in particular on fish populations.

Therefore hydraulic, economic as well as ecological aspects have to be considered when assessing the suitability of a potential flood polder location. These aspects were investigated at a proposed flood polder at the Middle Elbe River, Germany. The flood polder is designed for reducing flood peaks having a 100 years recurrence interval or more. The maximum volume of the storage amounts to 44 million m^3. The paper summarizes research results from (I) the hydrodynamic modelling, (II) the water quality modelling and (III) the estimation of annual losses to be expected in the flood polder.

(I) The hydrodynamic modelling is based on a 1D and a coupled 1D-2D approach for the flood polder and the adjacent river stretch. Beyond the simulated flood peak attenuation, the latter approach enables a two-dimensional representation of the flooding process within the flood polder in order to derive inundation parameters such as water depth and flow velocity as input for the subsequent vulnerability analysis. (II) Based on the hydrodynamic model set-up, the water quality modelling focuses on the oxygen balance of the flood polder and the river stretch. Due to the poor data availability a rather simple approach with a low number of state variables including dissolved oxygen, biological oxygen demand and temperature was applied. (III) The damage estimation was carried for grassland, arable land and roads according to the prevailing land use types in the flood polder. Different from road damage, losses on agricultural fields exhibit a strong seasonal pattern, while the flooding probability also has a seasonal variation. Both seasonal aspects were combined in the estimation of annual losses on agricultural fields.

Finally, general conclusions are drawn from the presented case study. They include recommendations on the choice of location, operation strategy and land use of flood polders.

Keywords: Flood polders, flood risk, hydrodynamic modelling, water quality modelling, damage estimation, Elbe River

Please consider this abstract for the Young FLOODsite prize.

Development of estuary morphology models

J.M. Huthnance & A. Lane
NERC Proudman Oceanographic Laboratory, Liverpool, UK

H. Karunarathna
Department of Civil Engineering, University of Glasgow, Glasgow, UK

A.J. Manning
School of Earth, Ocean & Environmental Sciences, University of Plymouth, Plymouth, UK

D.E. Reeve
School of Engineering, University of Plymouth, Plymouth, UK

P.A. Norton & A.P. Wright
ABP Marine Environmental Research, Southampton, UK

R.L. Soulsby, J. Spearman & I.H. Townend
HR Wallingford, Wallingford, UK

S. Surendran
Environment Agency, Reading, UK

Estuary management is a key part of flood and coastal erosion management. Management of estuarine morphology and associated flood risks needs well-validated predictive models, which have been lacking. The UK Estuaries Research Programme (ERP) addresses the challenge of predicting sediment dynamics and longer-term changes in morphology. Project FD2107 in ERP aimed to enhance "Hybrid" models combining elements and advantages of "top-down" and process-based "bottom-up" approaches. Developments include:

- An *Analytical Emulator*, based on 1-D hydrodynamics with constant uniform side-slope. It indicates total area/volume response to water levels and tidal range with minimal computation. However, its form limits realism in representing high- and low-water areas;
- A *Hybrid Regime* model, combining 1-D hydrodynamics with *regime* relations between discharge, cross-section area and width. For changed conditions, individual cross-section changes are predicted, constrained by solid surfaces and the overall regime relations;
- *Morpho-SandTrack*, enhancing a pre-existing model *SandTrack* with 2-D hydrodynamics, sediment dynamics and particle-tracking. The enhanced model predicts changing morphology; the flow model is re-run as bathymetry evolves;
- A *Realignment* model, predicting local changes in morphology and saltmarsh due to managed realignment. The model uses 2-D hydrodynamics (flow and waves) to spread sediment in a localised area;
- *ASMITA*, a pre-existing model now programmed in Matlab. ASMITA evolves the size of aggregated intertidal area, channels and ebb-tidal delta; these exchange sediment tending to fill accommodation space;
- An *Inverse* model evolving depth in 2-D according to a diffusion equation plus "source" derived from past depth changes. Prediction is limited: to auto-correlation times of most the "source"; to scenarios with precedents.

These models, and another particle-tracking model, were applied to eight varied UK estuaries. Scenarios were run to identify impacts and sensitivities (e.g. to sea level and sediment properties).

Results suggest that intertidal area usually decreases as sea level rises. Predicted sediment supply would usually enable infill keeping pace with sea-level rise, but models differ in whether such infill occurs. Small changes in tidal range have small effect. More river flow (+20 per cent) gives mostly small changes. Different estuaries respond differently, calling for specific predictions using models.

These applications illustrated models' respective merits. Models should be validated against historic change or by intercomparison. FD2107 has also provided inferences for flood risk, defences and habitats. The FutureCoast database has been augmented.

Keywords: Flood risk management, morphology, modelling

A GIS-based risk assessment methodology for flood pollutants

A. Sauer, J. Schanze & U. Walz
Leibniz Institute of Ecological and Regional Development (IOER), Dresden, Germany

Research on the consequences of flood events has so far focused on damages to buildings and infrastructure caused by water. Only little interest has been payed to pollutant impacts on human health and the ennvironment. Toxic substances such as trace elements (e.g. As, Cd, Hg, Pb, Zn) and persistent organic pollutants (e.g. HCH, DDX) released from contaminated river bank sediments or former industrial sites are transported by the flood water and accumulate in the floodplain, mainly bound to fine particles. Depending on the land use in the floodplain, different receptors like people, cattle, and food and fodder plants could be affected by the contaminants. The holistic management of flood risks therefore needs to integrate procedures for assessing the multiple risks caused by different substances influencing different receptors. At present, comprehensive approaches for the analysis, assessment and reduction of multiple contaminant risks due to floods—notably taking the spatial dimension into consideration—are still missing.

The paper presents an integrated GIS-based risk assessment methodology for flood pollutants, consisting of the sound combination of methods from flood risk analysis and contaminant risk assessment, using an adopted SPRC approach as methodological frame. The work is embedded in the SARISK research project: "Development of a pollutant fate and transport model for the risk analysis and assessment of contaminants due to extreme flood events". The approach of SARISK consists of hydraulic modelling and matter-flux propagation based on pre-calculated flood waves and flood frequencies (flooding scenarios). The applied models provide 2D-inundation simulations followed by sedimentation and coupled fate and transport simulations of selected substances, resulting in maps of the deposition probability of the substances.

Due to the outstanding relevance of the receptor "human", the focus is set on the human health risk assessment, comprising the following elements:

- At first, the hazard analysis describes the release, transport and accumulation of substances based on flood characteristics and substance properties as well as fate and transfer processes after sedimentation (e.g. soil-to-plant transfer). It covers the identification of pathways from the sources to the interim deposition. Results are maps of spatially distributed substance concentrations in environmental sediments, i.e. the contaminant source.
- Secondly, the exposure analysis performs the linkages between the contaminant sources to the receptors via the pathways (e.g. soil ingestion, food consumption). It consists of a receptor analysis which indicates and characterises potential human receptors derived from land-use types, taking their spatial distribution and properties (e.g. age, activity budgets) into account. Subsequently, the exposure analysis quantifies the exposition of the receptors to a certain substance by calculating transfer and intake rates. The results are maps displaying amounts of absorbed contaminants for a certain receptor, e.g. Daily Intake Rates (DIR).
- Finally, the contaminant risk determination describes the effects (consequences) on the receptors by dose-response relationships, forming the basis for the final substance-based risk assessment. This is carried out by comparison of the calculated exposures (DIR) with toxicological reference values, e.g. Tolerable Daily Intakes (TDIR), resulting in maps of Risk Indices for each receptor.

Keywords: Flood risk management, risk assessment, exposure analysis, flood pollutants, trace elements, persistent organic pollutants

Environmental impact of flash floods in Hungary

Szabolcs Czigány, Ervin Pirkhoffer & István Geresdi
*Department of Soil Sciences and Climatology, Institute of Environmental Sciences,
University of Pécs, Pécs, Hungary*

Until today 39 documented flash floods have been reported from Hungary, causing considerable economic loss and, in certain cases, fatalities. Flash floods can potentially occur along a total stream length of 1300 km and threaten several settlements along stream segments. Based on topography, soil properties, land use and the locations of documented flash floods, areas of high flash flood risk cover a total watershed area of 5750 km^2 in Hungary that comprises about 16% of the hilly and mountainous land area of the country.

To prevent, or at least mitigate, extensive economic loss and fatalities, the "Forecast of intense convective rainfall events and study of their environmental" nationwide environmental project (supported by the Jedlik Ányos foundation) aims to create a nationwide flash flood guidance (FFG) system for the hilly and mountainous regions of Hungary (approx. 36,000 km^2). The major objectives of the present study are the followings:

1. to analyze the degree of correlation between the localities of documented flash floods and (i) rainfall intensity and (ii) rainfall amount per rainfall event,
2. to determine spatial correlation between localities of documented flash floods and other environmental factors (e.g. canopy cover, soil and rock type) that may enhance flash floods,
3. to delineate the most flash flood affected watersheds in Hungary,
4. to determine the runoff threshold values that triggers flash floods for study watersheds in the Western Mecsek Mountains in SW Hungary by using the HEC-HMS hydrological modeling software,
5. to create a nationwide flash flood guidance system (FFG) based on a flowchart model and local discharge precipitation gages.

Analytical tools and procedure include (i) numerical modeling of flash floods, (ii) field studies and (iii) GIS application.

Precipitation intensities and total measured amounts per rainfall events showed a poor spatial correlation with the documented and reported flash flood localities. Consequently we need to conclude, that solely based on a long-term (30-year) precipitation data, forecasting of flash floods is challenging and hazardous. To delineate flash flood-affected watersheds, environmental factors other than precipitation need to be considered. Based on the 10-meter resolution digital elevation model of Hungary, the Corine Land Cover land use database and rock type, correlations between documented flash flood localities and (i) slope, (ii) ratio of forest cover and (iii) rock type was found.

In a selected study watershed (Bükkösd Valley) in the Western Mecsek Hills (SW Hungary) we determined the threshold runoff values that represent various environmental scenarios. Threshold runoff values are based on season, soil (moisture) saturation, ABR intensity and duration, and presence or absence of snow pack. Threshold runoff values were determined with the HEC-HMS hydrological modeling software by using the 15-minute resolution discharge and 10-minute resolution average basin rainfall (ABR) data of a long term monitoring system. The scenarios are organized into a flow chart that forms the base of the local FFG system that later will be further expanded for all flash flood-prone watersheds of Hungary. The Bükkösd Valley study area is also suitable for model data validation and model calibration.

Keywords: average basin rainfall, flash flood guidance, hydrological modelling, Hungary

Predicting beach morphology as part of flood risk assessment

J.M. Horrillo-Caraballo & D.E. Reeve
School of Engineering, Centre for Coastal Dynamics and Engineering (C-CoDE), Plymouth, UK

In recent years, as a result of the implementation of a strategic approach to flood and coastal management, the necessity for more robust methodologies for incorporating risk assessments within coastal engineering design has been clearly identified. Climate change will modify wave and water level conditions and for this reason affect the vulnerability of coastal defence structures. There are a large number of coastal structures whose stability depends upon the characteristics of the beach where they are located. From this perspective, it is necessary to understand how seawalls and beaches interact during storms and how beach profiles interact with waves, tides and storms. Without this, it is difficult to provide realistic estimates of interaction between the structure and the beach profiles and how they are likely to change over time. Sea walls are often part of a larger scheme, operating in combination with beach and other structures (e.g. beach nourishment, groynes, etc.) to control wave energy, thereby improving the resistance to coastal erosion or limiting wave overtopping.

In order to evaluate the importance of interaction between waves, storm surges and profile response, long-term bathymetric surveys at a coastal segment of the English south coast have been investigated with Canonical Correlation Analysis (CCA).

This method have been used in different fields and especially in the coastal field to detect patterns in the changes of beach profiles and prevailing wave conditions. Here, we analyse historical data obtained along the frontage of Milford-on-Sea. The beaches along this segment of coast are mainly composite, with a shingle upper beach and sandy lower beach, or mixed sand and shigle.

CCA is used first to analyse historical records to identify linkages between beach profile changes and wave conditions. It is then used in a forecasting mode as a tool for prediction. The regression matrices derived from the data sets on profiles and waves were used to reconstruct the time series of profiles using a limited number of CCA modes, which were then extrapolated for predictions.

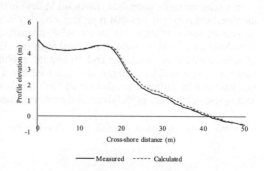

Figure 1. (a) Seawall at Milford on Sea and (b) measured and calculated profiles with offshore wave conditions (date 23/05/2006).

Figure 1b shows the measured beach profile for May 23th/2006 and the prediction made from January 31st/2006. Satisfactory agreement is obtained in the area where the profile shows evidence of considerable change, whereas in the offshore the error is well assimilated by the prediction.

In the full paper we will present a detailed study of the relationship between waves, storm surges and profile response. This relationship could be used by coastal planners to predict profile evolution based on the waves and storm surges. The CCA approach provides a method for forecasting profile response using offshore and nearshore wave properties.

Keywords: Beach profiles, Canonical Correlation Analysis, forecasting, profile response

Alkborough scheme reduces extreme water levels in the Humber Estuary and creates new habitat

D. Wheeler
Halcrow Group Ltd, Hampton, Peterborough, UK

S. Tan & N. Pontee
Halcrow Group, Swindon, UK

J. Pygott
Environment Agency, Leeds, UK

There are a number of large scale managed realignment schemes that have been undertaken across Europe under the umbrella of the FRaME project. Alkborough is one such scheme, situated on the south bank of the Humber on the eastern side of the confluence between the River Humber and River Trent. The site is bounded to the south east by the escarpment of the Lincolnshire Heights with the remaining boundaries of the site being formed by the shoreline of the estuary, approximately 6km in length. The site is principally in the ownership of Natural England and the Environment Agency.

The promotion of the Alkborough scheme was driven by strategic estuary wide requirements and has significant benefits with regard to sustainable estuary management, namely by:

- Providing flood storage to reduce peak tide levels in the inner estuary and tidal rivers, thereby deferring the need for work elsewhere in the estuary to address sea level rise; and
- Contributing to habitat creation responsibilities under the European Union Wild Birds and Habitats Directives by creating 170 ha of new inter-tidal habitat and 200 ha of other natural habitats.

This paper describes some of the work that was undertaken prior to construction starting on site, in particular:

- Hydrodynamic modelling and geomorphological analysis to inform the assessment of the external impacts of the development and ensure that these were alleviated by appropriate design;
- Economic assessment; and
- Environmental impact assessment.

The key environmental issue with respect to the design of the scheme was the potential tide level reductions in the estuary and the effects of this on navigational interests. As a result normal tidal inundation of the site is by way of a fixed weir breach structure. This structure restricts the inundation into the site to volumes that have negligible effect on tide levels in the estuary. The construction of an overspill weir by lowering part of the existing tidal defence permits floodwaters to enter the site on surge tides within the estuary.

The scheme demonstrates how strategic reductions in water level can be made within an estuarine environment whilst simultaneously creating new areas of inter-tidal and other habitats. It is anticipated that the habitat and species supported by the scheme will, in time, allow the site to be designated as a National Nature Reserve (subject to designation by Natural England's Council) and may ultimately allow it to be included within the European Site boundary.

Keywords: Flood risk management, habitat creation

Managing coastal change: Walberswick to Dunwich

M. Cali, A. Parsons, N. Pontee, L. Batty & S. Duggan
Halcrow Group Ltd, UK

P. Miller
Environment Agency, UK

The Government has acknowledged that, where flood management measures are essential to maintain designated sites of European importance (European sites) in a favourable condition, it is legally obliged, through the Conservation (Natural Habitats &c.) Regulations 1994 for England and Wales, to provide these measures.

This paper focuses on the appraisal of options for managing the coastal change that is occurring on the English east coast in Suffolk between Walberswick and Dunwich. The shingle barrier and marshes at this site are of international conservation interest and form part of the Minsmere to Walberswick Special Protection Area (SPA) and Special Area of Conservation (SAC).

Historically, the shingle barrier, a conservation feature itself, provided a flood risk reduction function to the area behind the barrier. This allowed the development of freshwater marshes and reed beds of considerable environmental interest, resulting in designation of the protected area as SPA. However, the shingle barrier, which is a key feature of the SAC, has a history of breaching during storm events and in recent years breaching has become an annual or greater occurrence.

In the past, the Environment Agency has managed the flood risk to the environmental sites, and to people and property in the villages of Dunwich and Walberswick, through repairing the breaches and maintaining the crest level of the shingle barrier using material from the foreshore. This practice is now considered to be unsustainable due to lack of natural shingle being deposited on the beach, and the damage that the repair activities cause to the interest features of the shingle ridge itself.

The Environment Agency has strategically considered flood risk management options for the whole flood cell. A key consideration was the requirement to protect or replace the freshwater habitats, resulting in the need to manage the adaptation of the site to increasing tidal flood risks and to allow time to provide replacement freshwater habitat.

The paper will describe the sometimes conflicting policy issues that have been considered for this site, and disseminate lessons that should help develop best practice in the appraisal of schemes required to manage adaptation to coastal change.

Keywords: Shingle, adaptation, coastal change, European Sites

Uncertainties in the parameterisation of rainfall-runoff-models to quantify land-use effects in flood risk assessment

A. Wahren, K.H. Feger, H. Frenzel & K. Schwärzel
Institute of Soil Science and Site Ecology, Dresden University of Technology, Germany

The European Flood-Directive points out the need of flood risk maps. Such maps will be the basis of flood risk management plans focussing on prevention, protection, and preparedness. This work has to be done river-basin-oriented and directly linked to the EU-Water-Framework-Directive. Thus, emphasis has to be put on the detection of synergy effects between the good ecological status (e.g. minimize technical impacts) of the water bodies and flood protection.

The revitalization and increase of natural water retention potentials—notably in headwater catchments—is a major component of sustainable flood prevention strategies. It is logical that changes in land-use patterns (e.g. expansion of settlements including road-construction, deforestation, distinct practices in arable and grassland management) contribute to an increased frequency and severity of flood generation. For forest land-use, it has been stated that afforestation and the promotion of close-to-nature silviculture will considerably increase the water retention in landscapes. However, there is a controversial debate on the quantitative role of such non-structural flood risk management measures with respect to event size and scale-based physical conditions. Modelling approaches very often neglect important aspects when rainfall-runoff-models are parameterised. Hence, many models just consider vegetation parameters (root depth, leaf area index (LAI), canopy hight, etc.). Some more advanced models (AKWA-M®, WASIM-ETH, etc.) also include pre-event soil water content by calculating land-use specific evapotranspiration. In addition, one should also be aware that changes in the vegetation cover (e.g. conversion from arable land into grassland or forest) in the mid- to long-term will also result in distinct changes in soil hydraulic properties (infiltration, percolation, retention).

Our investigation in the catchment of the upper Mulde (Saxony/Germany, FLOODsite: Elbe pilot-area) explored the impact of afforestation measures on the soil hydraulic properties. 'False chronosequences' were used to quantify the time-dependent dynamical character of such changes. Four plots were identified at a test area with comparable pedological start conditions and a set of tree stands of different age: (1) arable field (initial state); (2) 6-y-old afforestation; (3) 50-y-old afforestation; (4) ancient natural forest ('target' stocking). Water retention curves and unsaturated conductivities were analysed in the lab and infiltration capacities were measured in the field. Distinct differences were detected. Especially the top soil layers showed an increased conductivity and a higher portion of coarse/middle pores causing an increased infiltration and soil water retention potential (Table 1).

Table 2 shows an example from sites in the Mulde catchment (Zellwald). The model (AKWA-M®) simulates the runoff behaviour influenced by afforestation for different pre-event soil moisture conditions. The runoff peaks were calculated comparing two model parameterisations ("soil properties unchanged" and "soil properties changed") for two flood generating rainfall events. If the changed soil hydraulic behavior is neglected, the

Table 1. Pore distribution [Vol%], related field capacity (mm) and unsaturated hydraulic conductivity [mm d^{-1}] in the top layers (30 cm).

Land-use	Horizon	Pore diameter [μm]			Field Capacity [mm]	Hydraulic Conductivity [mm d^{-1}] at pF 2,5
		>50	50–10	10–0,2		
Arable land	Ap	6	5	3	106	0,11 ± 0,02
Young afforestation	Ah	6	9	7	149	0,26 ± 0,05
Old afforestation	Ah	5	14	10	179	0,43 ± 0,10

Table 2: Peak runoff from an afforested site for initial state and two model parameterisations ("soil pro-perties unchanged" and "soil properties changed") for the 'target state' land-use (old forest).

	P (T = 25 a, D = 2h) [mm]	HQ(25) [mm]			Reduction [%]				P (T = 100 a, D = 2h) [mm]	HQ(25) [mm]			Reduction [%]		
		Pre-event soil moisture			Pre-event soil moisture					Pre-event soil moisture			Pre-event soil moisture		
		high	middle	low	high	middle	low			high	middle	low	high	middle	low
Arable Land	45	30	27	17	–	–44	–45	56		41	39	27	–	–37	–39
Forest ('target state')		17	14	12	–44	–45	–29	Soil properties unchanged		26	24	22	–37	–39	–19
		15	13	11	–50	–50	–37	Soil properties changed		24	22	21	–41	–43	–23
Difference					6	5	8						4	4	4

251

peak reduction for HQ (25 a) and HQ (100 a) is smaller (4–8%) than in the other case. The reason for that difference is the increased conductivity causing a higher infiltration, and the increased field capacity which enables the soil to store more water.

Our study underlines that rainfall-runoff models should consider changes both in vegetation and soil properties. Only such a combined approach ensures to address land-use effects in an appropriate way. The lack of relevant data on changes in soil properties should not lead to the conclusion, that land-use measures are ineffective. It is quite clear that their impact is limited, but the sustainability and the synergy effects to nature protection and soil conservation should keep the considerations about that non-structural measures vital.

Targeted measures to improve the natural retention in the watershed are long-term challenges and, thus, the appearance of the benefits may take decades. As a consequence, reliable model calculations are inevitable to estimate the actual potential of land-use strategies and their limitations. Beside the use of state-of-the-art models to predict effects for the present planning and building of flood protection measures, also tools to predict key parameters should be improved continuously in order to minimise the uncertainties.

Keywords: Land-use, soil hydaulic properties, flood, model parameterisation, water retention, conductivity, infiltration

Impact of the barrage construction on the hydrodynamic process in the severn estuary using a 2D finite volume model

Junqiang Xia
Cardiff School of Engineering, Cardiff University, Cardiff, UK
State Key Laboratory of Hydroscience and Engineering, Tsinghua University, Beijing, China

Roger A. Falconer & Binliang Lin
Cardiff School of Engineering, Cardiff University, Cardiff, UK

The Severn Estuary has a tidal range of approaching 14m with the second highest in the world, and a barrage is proposing to be built for power generation. Therefore, the Severn Barrage will be a very important project in UK, and the impact of the barrage construction on the hydrodynamic process in the Bristol Channel and Severn Estuary should be accounted for beforehand. A two-dimensional hydrodynamic model based on an unstructured triangular mesh is introduced here, and this model uses a TVD finite volume method to solve 2D shallow water equations, and gives the second-order accuracy in time and space. The developed model for the Severn Estuary is calibrated firstly by observed tidal ranges and currents at different sites from an Admiralty Chart, and is verified further by the observed data in May 2003. Then, a technique of domain decomposition and a treatment of internal open boundary are presented in the model to account for the presence of different hydraulic structures along the Severn Barrage, and the impact of the Severn Barrage construction on the hydrodynamic process in the Severn Estuary is investigated using this model. Simulated hydrodynamic results with and without the barrage are analyzed in detail. Model predictions indicate that the mean power output can reach 2.0 GW over a typical cycle of mean spring tide, and the maximum discharge across a section near the M4 Bridge can reduce by 50 %, as compared with the value without the barrage, which makes the maximum water level upstream of the barrage decrease by 0.5–1.5 m due to the construction of the Severn Barrage. In addition, the complex velocity fields in the vicinity of the barrage are predicted for different operation modes.

Keywords: Severn Estuary, Severn Barrage, hydrodynamic process, two-dimensional model, finite volume method, unstructured triangular mesh, domain decomposition

Risk sharing, equity and social justice

From knowledge management to prevention strategies: The example of the tools developed by French insurers

J. Chemitte & R. Nussbaum
Mission risques naturels (MRN), Paris, France

Many factors justify the renewed interest of territorial stakeholders for a deeper reflection on their ability to deal with threats posed by natural events:

- economic: the rising costs of natural disasters, the threat of climate change,
- societal: the increasing risk aversion and human activities vulnerability,
- regulatory: the new European solvency standards, a future reform of the French insurance system against natural disasters,
- technological: geographic information systems, alternative risk transfer techniques.

It is indeed appropriate that organizations undertake a dynamic and pro-active way which provides them the means to achieve their goals or missions in all circumstances, particularly in cases of stress or surprise.

The first step in effective natural risk management requires organizations, including insurance companies and their customers, to engage the understanding on their level of exposure to different natural hazards. Therefore, the challenge of this strategic knowledge management is to find the right knowledge at the right time. This underlines the need to model and introduce that knowledge into a system to make it easily accessible. It involves mobilizing the organization's know-how, but also to integrate or to confront the resulting outputs with external expertise or other source of knowledge.

The Mission risques naturels association (MRN) has been created in 2000 by the French insurer's federations. Its missions serve the general interest of insurance companies and it plays a role as interface with other stakeholders in the French natural risks management scheme. In consequence, it is its responsibility to organize and manage the knowledge relating to natural risks, in order to contribute to the French prevention policies improvement.

The introduction will deal with the context of natural risks management in France and propose a quick overview of the tools, which are used by the different risk management stakeholders. Then the intervention will focus on the dynamic process of knowledge creation implemented by the MRN for the insurance industry, based on the creation and animation of:

- communities of practitioners, defined as groups of people who share a set of problems or a common interest on a subject and who deepen their knowledge and expertise in this area by interacting on a regular basis (Wenger, McDernmott et Snyder, 2002);
- geospatial collaborations, which refers to collaborative situations using geographic data and GIS technology ("work together using geographic data", Pornon and Noucher, 2007);

The next part will present the information system developed by the association (also known as geospatial information infrastructure). This tool is accessible from the Web site of the association to the employees of insurance companies. It delivers methods, products and services to meet the needs of several lines of business. An example of online services will be submitted in the case of a prevention engineer working for an insurance company and who has to survey a customer's site to assess the risks he wants to cover.

Finally, a few comments on the successful experience of deterministic portfolio exposure modelling realised by a company will open a discussion on the future tools for natural risks management in the organizations.

Keywords: Flood risk management, insurance, GIS, Spatial data infrastructure, geospatial web services

What's 'fair' about flood and coastal erosion risk management? A case study evaluation of policies and attitudes in England

C. Johnson, S. Tunstall, S. Priest, S. McCarthy & E. Penning-Rowsell
Flood Hazard Research Centre, Middlesex University, London, UK

The extent to which those at risk of flooding or coastal erosion regard governmental decision-making to be 'fair' or 'unfair' depends not only on whether the outcomes of governmental processes are regarded as 'fair' but also on whether the processes by which decisions are made are considered to be 'fair'. Using evidence from an ongoing Defra funded research project—evaluating the 'socially just' nature of flood and coastal erosion risk management (FCERM) policy and practice—this paper examines the policies and attitudes of key stakeholders towards 'fair' FCERM in England.

The research underpinning this paper examined the extent to which national policy and guidance embodies the three justice principles of equality, vulnerability and maximum utility as dominant values influencing decision making processes and outcomes. The findings illustrate both institutional differences in the application of social justice principles across government policy and departments and significant differences depending on the 'type' of adaptation option under consideration. For example, maximising utility is the dominant principle for the Treasury and Defra whilst there is more scope for vulnerability and equality principles in the policies and guidance of the Environment Agency and CLG. Similarly, policy and guidance for the prioritisation of flood defence options are, as expected, dominated by considerations for economic efficiency underpinned by maximum utility principles. Elsewhere in the flood risk management portfolio such utility considerations feature minimally, if at all: flood warnings, emergency response and spatial planning being three such examples. Thus, whilst advances have been made in national policy for recognising the importance of, and seeking to address, fairness issues in FCERM, there remains a lack of joined-up thinking across institutions, policy and guidance.

Attitudes to the 'fairness' of FCERM processes and outcomes tend to recognise the contested nature of the concept with different interpretations dependent on the stakeholders and context concerned. Similarly, the relatively recent and evolving introduction of 'fairness' as a key issue in government policy was highlighted as significant; where Making Space for Water is regarded as a 'real' policy drive to make FCERM 'fairer'. Given this general attitude, of the more specific issues that were raised in interview and discussion with key stakeholders, the following were considered particularly important in an evaluation of the 'fairness' of FCERM: funding arrangements (national versus local; beneficiary pays principle; external contributions); coasts versus rivers; urban versus rural; fairness between options (e.g. between defence, flood warnings, spatial planning, resistance and resilience measures); fairness between different types of flooding; coastal adaptation and inter-generational equity.

Given that fairness of outcome is inevitably contested and that FCERM will always be constrained by the levels of funding available—one of the key findings of the research reported on here is the significance of procedural justice; particularly important of which are decision processes that are transparent and understandable. This is largely due to the continuing conflict between the requirements for national consistency to ensure that all people are treated equally and the requirement for a bottom-up approach to decision-making which is genuinely participatory, engaging people actively and flexibly in the decision-making process. Managing this conflict is arguably one of the biggest challenges for 'fair' FCERM in England.

Keywords: Flood risk management, coastal erosion, social justice, fairness, England

Flood risk perceptions in the Dutch province of Zeeland: Does the public still support current policies?

Jörg Krywkow, Tanya Filatova & Anne van der Veen
University of Twente, Enschede, The Netherlands

The conventional method of risk analysis (with risk as a product of probability and consequences) does not allow for a pluralistic approach that includes the various risk perceptions of stakeholders or lay people within a community or region. In river basins, it is often an expert-based economic analysis of land use values that serves as the underlying estimation of costs of risk mitigation. Intangibles such as nature development, biodiversity and cultural heritage are difficult to include in this calculation. Yet, local or domain knowledge can be used to develop a realistic approach for including tangibles and intangibles in the assessment of land use values, as well as developing new approaches to risk management. This can be an important link in developing appropriate solutions for the mitigation of risk.

Risk perception is derived from the psychometric paradigm that distinguishes a variety of risk characteristics. Due to a number of similarities among these risk perceptions all known risk characteristics can be condensed into three higher order risk characteristics—*awareness*, *worry* and *preparedness*.

Risk perception information may be collected with the help of knowledge elicitation methods. These methods comprise interview, questionnaire and survey approaches that belong to the consultation level of participatory methods. The role of public and stakeholder participation is often underestimated, and not always wisely applied. Furthermore, experts such as planners, engineers and ecologists may have difficulty in accepting and incorporating domain knowledge in their projects. However, public and stakeholder opinion can help to improve the flood protection policy in a river basin or even at the level of province or country. European directives such as the Water Framework Directive and the Flood Risk Directive emphasize the active participation of involved people and organisations in policy processes.

On the Dutch side of the Scheld estuary, the authors conducted a number of personal interviews, which were followed up by a survey sent to 3000 households within the province of Zeeland (The Netherlands). The questions addressed flood perceptions as well as opinions about current and desired flood policy, evacuation and early warning systems and the perceived role of responsible authorities.

The flood perception approach used here assumes a particular level of 'voluntariness' where risk outweighs benefit within a flood-prone area. The example of the Ebro delta (Spain) demonstrates that stakeholders may be able to decide by themselves whether or not to invest in flood protection measures such as land-use change. This delta shore is not generally protected. In contrast, the Dutch coastal area is completely protected by dykes. Most of the so-called polders are low-lying lands between 0 and about 6 meters below sea level. This implies well-organised water management authorities ('Waterschappen', 'Rijkswaterstaat'), who are able to maintain the flood protection of the low-lying areas. Individual trade-offs between risk and benefit of flood protection is virtually not possible. However, the implementation of auxiliary measures such as compartments, additional flood plains, evacuation plans, nature protection and similar responses may be issues for discussion among the public, stakeholders, water managers and policy makers.

Despite reduced levels of voluntariness in the polders, the variety of public opinions is remarkable, and helps to enrich or strengthen current flood protection policy.

Keywords: Risk perception, psychometric paradigm, knowledge elicitation, risk-benefit trade-off, participatory process

A partnership approach – public flood risk management and private insurance

Matt Crossman
Department for Environment Food and Rural Affairs, London, UK

Swenja Surminski & Arthur Philp
Association of British Insurers, London, UK

Dan Skerten
Department for Environment Food and Rural Affairs, London, UK

Flood risk poses a significant threat to many communities and the effects of climate change, particularly sea level rise, will lead to significant increases in the level of risk. Whilst traditional engineering approaches can reduce the likelihood of damage, and techniques are increasingly being deployed to minimise impacts in the event of flooding, they can never entirely eliminate the risk. This paper will provide a description of the joint review of the Statement of Principles undertaken following the major floods in England during summer 2007. It will set out the renewed partnership in the United Kingdom that enables private insurance, to be made widely available on the basis of Government commitment to manage risks and contrast this with alternative approaches.

Government has announced that funding for flood and coastal erosion risk management in England will increase from approximately £600 million in 2006/7 to £800 million in 2010/11, but recognises that insurance and other tools are essential to allow those living with the residual risk to cope. The partnership approach to flood risk management adopted in the UK is different from that in many parts of Europe and elsewhere. Table 1 summarises a review by Comité Européen des Assurances of the coverage of natural perils for property insurance in European markets.

Functioning partnerships between the government and private sector are limited to a relatively few countries. In locations where cover is optional and not widely taken up adverse selection may cause problems. This is where only those most at risk choose to insure and the arrangements become either very expensive or unsustainable. Where flood cover is not generally available or penetration is low, the consequences in the event of a flood will be borne either by those directly affected or, more likely, by the nation as a whole. In these countries, as well as those where

Table 1. Type of flood insurance and rate of penetration for European countries.

	Penetration of flood risk cover			
	<10%	10–25%	25–75%	≥75%
Compulsory cover				Belgium* France Switzerland
Obligatory pool			Spain	Norway
Optional cover	Germany Greece	Austria Finland Italy	Czech Republic Poland Portugal	Sweden Turkey United Kingdom

*penetration data not available but legal requirement introduced in September 2005.

Keywords: Flood risk management, Insurance, partnership

compulsory cover is not based on risk, government and taxpayers effectively becomes responsible by default, often compensating those affected. This penalises those who address their risk by making them subsidise those who don't, introducing economic inefficiency and failing to signal the true costs of occupying high risk locations.

Work on the review of the Statement of Principles is currently underway, but is expected to be completed in the summer.

The international teaching module FLOOD*master* – an integrated part of a European educational platform on flood risk management

Jörg Seegert, Christian Bernhofer & Katja Siemens
Technische Universität Dresden, Institute of Hydrology and Meteorology, Dresden, Germany

Jochen Schanze
Leibniz Institute of Ecological and Regional Development, Dresden, Germany

Knowledge transfer and training activities have become more and more important in the field of flood risk management not only with respect to preparedness and warning of the affected people but also in terms of a trans-disciplinary education of students, graduates, and professionals. For a long time neither in Germany nor in Europe a university training didn't exist exclusively focusing on the issues of integrated flood risk management. Therefore, one major task of the European Integrated Project FLOOD*site* on "Integrated Flood Risk Analysis and Management Methodologies" was to develop a European master course package on Flood Risk Management (FLOOD*site* European Master—*FEM* platform). This platform was first initialized and integrated by two existing master courses: 'Hydrogeological Risk Mitigation' at University of Padova and 'Hydro Science and Engineering' with a clear focus on Flood Risk Management (FLOOD*master*) at Technische Universität Dresden. Both are consistent with the educational scheme following the Bologna Treaty of the EU and establish a sound basis for a new European Master Programme on flood risk management linking courses in Germany, the Netherlands, and Great Britain.

The FLOOD*master* course at TU Dresden already started in the summer term 2005 and provides comprehensive knowledge on integrated flood risk management based on a theoretical framework, disciplinary contributions, and real-world requirements. It encompasses theories and methods from natural and engineering sciences as well as from economic, social, and planning sciences, and allows participation on different individual backgrounds. Primary target group are graduate students of different master programmes and graduates in the 4th/5th year (esp. from, hydrology, water management, hydraulic engineering, geography etc.) as well as professionals. Main topics are various types of flood hazards, areas of vulnerability, risk and risk reduction as well as management strategies. All aspects are integrated in the study concept based on two classes considering the whole cascade of risk generation from precipitation to societal consequences and all tasks of integrated flood risk management, three workshops on the different types of floods as well as an actors' workshop, a study tour to an international river basin, and a final study project. On the average, the course events are attended by 20 to 30 students, up to now about two dozen students successfully reached their FLOOD*master* certificate.

The concept of the study programme is prepared for both university study and distance learning courses. The teaching material is web-based for download and interactive use as study material for university lectures as well as e-learning conditions. This is a requirement of an open European educational platform offered to graduates and professionals of national and international universities, research institutions, administration, and consultancies. This facilitates options for inter-linking master courses all-over Europe. The development of the study programme was advised by a scientific committee and is closely linked to the national BMBF research programme RIMAX on "Risk Management of Extreme Flood Events" and international (esp. the EU-IP FLOOD*site*) partners. The module currently incorporates more than a dozen guest lecturers from institutions all over Germany and Europe.—The project is funded by the German Federal Ministry of Education and Research (BMBF promotional reference 0330680) within the research programme RIMAX and supported by the EU-IP FLOOD*site*.

Keywords: integrated flood risk management, education, knowledge transfer, university training, distance learning

Decision support for strategic flood risk planning – a generic conceptual model

A.G.J. Dale & M.V.T. Roberts
Capita Symonds Ltd, West Sussex, UK

Public reaction to the flood events in the UK, during the summer of 2007, clearly indicated the level of surprise felt by the community with respect to the extent and severity of flooding that occurred. The expectation was that the level of resilience should have been greater and the consequences less devastating. Alarmingly, a simple review of readily available data sets on flood risk would have identified that many of the consequences, observed during the events in June and July 2007, were completely predictable.

In practice, the actual flood risk management response to the flooding witnessed in July 2007, was led by the emergency services. Although the responses were carried out in a highly professional, efficient and committed manner, the required allocation of resources to specific consequences, such as single point failure of essential infrastructure, is concerning. Particularly, since it is predicted that climate change effects will increase the severity and frequency of emergency events. Whilst it is possible to identify the shortcomings in the existing roles, responsibilities or protocols adopted in respect to flood emergencies, as described in the interim outcome of the Pitt Review, there is perhaps a more fundamental need to understand why flood risk appears to be a topic that has particularly challenged those responsible for strategic responses.

This paper initially outlines common perceptions of flood risk and identifies fundamental areas of ambiguity or conflict that potentially reduces the clarity of, and confidence in, the decision support process. Consideration is then given to the possibility of identifying a common, consistent, coherent conceptual platform that would be more accessible to all parties, requiring an integrated model that can be applied at all levels of decision making; before, during and after a flood emergency.

A case study drawn from analysis of the Flooding in Gloucester in the summer of 2007 is examined and demonstrates the benefits that can be derived by application of simple, conceptual models to ensure critical essential utility infrastructure remains operational during flood events.

Who benefits from flood management policies?

Nigel Walmsley
HR Wallingford Ltd., Wallingford, Oxfordshire, UK

Edmund Penning-Rowsell
Flood Hazard Research Centre, Enfield Middlesex, UK

John Chatterton
John Chatterton & Associates, Birmingham, UK

Karl Hardy
Defra, London, UK

Large parts of England are at risk of flooding from rivers and the sea. Areas particularly at risk include the Humber corridor, the coastal areas in the South and East, low lying areas in East Anglia and the South West and major estuaries. Some 2.1 million properties are estimated to be in flood risk areas, affecting 4.3 million people (8.7 per cent of the population) and of these, around 469,000 properties are at significant risk of flooding (affecting 900,000 people).

Defra has policy responsibility for flood and coastal erosion risk management (FCERM) in England and the over-arching policy is *to reduce risks to people, property and the environment from flooding and coastal erosion through the provision of defences, flood forecasting and warning systems, increased flood resilience of property, beneficial land management changes and discouragement of inappropriate development in areas at risk of flooding*. Managing flood risk therefore encompasses a broad range measures but is not restricted to public intervention alone and factors such as flood insurance are also important within a risk management context.

FCERM measures provide a complex mix of public and private benefits to, and burdens on, society over long time periods. However, there is currently only limited understanding and evidence of how different interest groups and sectors benefit from public investments (or decisions not to invest) in reducing flood and coastal erosion risks. A series of case study assessments have been undertaken to provide evidence of the size and scale of benefits, costs and other intangible resource flows between different sector and interest groups. The case studies cover a broad mix of FCERM activities including the provision of defences, flood forecasting and warning, emergency response, flood resilience, spatial planning and flood insurance. The case study findings improve understanding of winners and losers, help to identify distributional imbalances, and can serve to inform future FCERM policies and programmes.

The research was commissioned as part of *Theme 1: Strategy and Policy Development* of the joint Defra/EA Research and Development (R&D) programme.

Keywords: Flood risk management, Defra/EA R&D

Uncertainty

Long term planning – robust strategic decision making in the face of gross uncertainty (tools and application to the Thames)

C. Mc Gahey & P.B. Sayers
HR Wallingford Ltd, Wallingford, Oxfordshire, UK

Robust strategic planning presents specific problems. Changing climates, changing socio-economic context as well as deterioration in defence assets all present the decision maker with complex policy choices with regard to what to do for the best.

This paper presents research undertaken in Task 14/18/24 of Floodsite and describes the development next generation decision support tools that support the user in developing meaningful strategic alternatives (based on a pipeline of multi-staged actions) and assessing their performance through time using credible system risk models (based on the so called RASP model) together with a structured assessment of their robustness based on their performance across a continuous space of climate and socio-economic futures (a concept that provides a significant step forwards compared to the discrete scenario analysis included in projects such as the Foresight future Flooding Project in 2004).

The science and tools developed through Floodsite is shown through application to the Thames Estuary. This is includes the development of strategic alternatives that include both structural and non-structural measures, climate change that includes changes in both fluvial and surge conditions, socio-economic that includes spatially varying population growth. The change in the defence infrastructure, deterioration and improvement, are also included together with more significant engineering interventions such as barriers and realignment.

The selection of robust decisions in the context of these changes is then explored, including the identification of non-regret solutions and flexibility decision pathways.

Keywords: Flood risk, uncertainty, long term planning, RASP

Anticipatory water management for advanced flood control

S.J. van Andel
UNESCO-IHE, Institute for Water Education, Delft, The Netherlands

A.H. Lobbrecht
UNESCO-IHE, Institute for Water Education, Delft, The Netherlands
HydroLogic BV, Delft, The Netherlands

R.K. Price
UNESCO-IHE, Institute for Water Education, Delft, The Netherlands

Anticipatory Water Management (AWM) is the operational management of water systems on the basis of forecasts of critical events. While in flood forecasting and real-time control measurements of water levels and precipitation play an important role, in anticipatory water management and early warning decisions have to be made before an event occurs. Therefore, the use of medium range meteorological forecasts is at the core of this research. The concerned application is flood control, hence the focus is on rainfall forecasts.

Ensemble rainfall forecasts provide forecasts of forecast uncertainty. This information allows for a wide range of risk based decision rules, to apply anticipatory control actions and how. For every particular water system and for every application, first a customised verification analysis of the ensemble rainfall forecasts must be done to assess the potential of Anticipatory Water Management and identify suitable decision rules. With ensemble rainfall forecasts, often these rules concern thresholds of rainfall (event thresholds) and probability. Also, a wide range of forecast horizons is increasingly available, meaning that a selection has to be made for consideration in daily control decisions.

A second important instrument in the application of AWM is water-system control modelling. In such modelling exercises next to the hydrological and hydrodynamic processes, also the (human-based) control of regulating structures needs to be modelled. This allows preparing water level forecasts on the basis of rainfall forecasts and governing control strategies. If these forecasts indicate flood levels, this can be seen as a warning that additional, anticipatory control actions are needed to reduce water levels. Using water-system control modelling, the operational water management can be simulated, allowing for new decision rules and control strategies to be verified in hindcast analyses.

In selecting new control strategies and in deciding on policy adoption of anticipatory control, water authorities need cost-benefit information. Benefits of reducing flood damage have to outweigh adverse effects of unnecessary anticipatory control actions as a consequence of false alarms. Adverse effects from unnecessary evacuations, controlled inundation, and low reservoir levels result in both tangible and intangible costs like damage to crops and loss of confidence in the warning system.

The framework for setting up Anticipatory Water Management will be demonstrated for a case study in the Netherlands. In this case-study the control of a channelled reservoir system needs to be optimised for flood reduction. Results of forecast verification and cost-benefit analyses will be presented.

Keywords: Ensemble forecasting, early warning, flood control, risk management, decision-making

Staged uncertainty and sensitivity analysis within flood risk analysis

Ben Gouldby & Greer Kingston
HR Wallingford, Wallingford, Oxfordshire, UK

Modelling flood risk is complex and associated with many sources of uncertainty. Models that are unable to capture the full physical processes that they are intended to represent are widely used and some physical processes, like breach formation, for example, are poorly understood. Furthermore, statistical modelling of extreme events is often based on relatively short periods of observed data and knowledge of basic parameters within the flood system, such as defence crest level or floodplain property, is subject to inaccuracies. Uncertainty analysis is intrinsically linked to flood risk analysis and it is increasingly becoming acknowledged as an important component to explicitly include within the decision making process. Whilst methods for uncertainty analysis have been available for many years, these typically become computationally intensive and impractical when applied to flood system risk analysis.

Variance based sensitivity analysis is a robust approach for assessing the contribution that individual input parameters and variables make to the output uncertainty of a model. This can be important for prioritising future data gathering exercises or model refinement research, where the variables contributing most to the uncertainty are prioritised for attention. In essence, this approach involves repeated uncertainty analyses to gauge the relative importance of a specific input variable in influencing the output uncertainty and can hence be extremely computationally demanding.

This paper details how the structure of a flood risk model can be reduced to a series of stages and how this can significantly reduce the computational burden of uncertainty and sensitivity analyses, yet still provide robust assessment. The methods are demonstrated on a case study.

Keywords: Flood risk, uncertainty, sensitivity

Assessing uncertainty in rainfall-runoff models: Application of data-driven models

D.L. Shrestha
UNESCO-IHE Institute for Water Education, Delft, The Netherlands

D.P. Solomatine
UNESCO-IHE Institute for Water Education, Delft, The Netherlands,
Water Resources Section, Delft University of Technology, The Netherlands

Flooding is a complex and inherently uncertain phenomenon. Consequently forecasts of it are inherently uncertain in nature due to various sources of uncertainty including model uncertainty, input uncertainty and parameter uncertainty. With rapid growths in computational power, increased availability of distributed hydrological observations and improved understanding of the physics and dynamics of the hydrological system allow the modeller/forecaster to build more complex and sophisticated flood forecasting models. While these advances supposed to improve the accuracy of the flood forecasting, they also increased the uncertainty due to the complexity of the models with increased parameters, amongst the other. It is therefore important to acknowledge the existence of the uncertainty in the flood forecasts and if possible every forecast should be complemented with the assessment of uncertainty. The uncertainty estimate in flood forecasting not only provides rational basis for flood warning and but also helps decision maker to use risk based decision, and thus it appraises the user of the uncertainty and enables the user to take risk explicitly into account.

A number of methods have been proposed in literature to estimate uncertainty in flood forecasting, e.g. simulation and sampling based, Bayesian and statistical, fuzzy set theory based methods and so on. The most of the uncertainty methods deal only with a single source of uncertainty, for instance, parameter uncertainty. This paper presents a novel method to estimate the total model uncertainty in flood forecasting using data driven modelling techniques. The method is referred to as UNcertainty Estimation based on Local Errors and Clustering (UNEEC). The UNEEC method is based on the analysis of the historical model errors that occurs in producing observed flows and computes total model uncertainty that takes into account all sources of uncertainty without attempting to disaggregate the contribution given by the individual sources of uncertainties.

In the UNEEC method, model errors (for example, mismatch between the observed and simulated values of runoff) are seen as the indicator of total model uncertainty. Since the direct analytical estimation of the probability distribution of the model errors is often difficult in the forecasting model, it is estimated separately for different hydrological situations using data driven models. The parameters characterizing these distributions are aggregated and used as output target values for building the training sets for the data driven models. This model, being trained, encapsulates the information about the model error localized for different hydrological conditions in the past, and is used to estimate the probability distribution of the model error for the new hydrological model runs. M5 Model trees are used as learning engines (however, other models like artificial neural networks can be used as well). Average mutual information and correlation analysis are used to determine the relevant parameters characterizing hydrological situations and the input variables for the learning models. The results are also compared with other uncertainty estimation approaches—GLUE and meta-Gaussian approach. The method is tested to estimate uncertainty of a conceptual rainfall-runoff model of Brue catchment in UK.

Keywords: flood forecasting, uncertainty analysis, data driven model, clustering analysis

Flash floods

European flash floods data collation and analysis

V. Bain & O. Newinger
CEREVE, École Nationale des Ponts et Chausées, Marne-la-Valée, France

E. Gaume
LCPC, Nantes, France

P. Bernardara
EDF, Paris, France

M. Barbuc
National Institute of Hydrology and Water Management, Romania

A. Bateman, J. Garcia, V. Medina, D. Sempere-Torres & D. Velasco
GRAHI, Universitat Politécnica de Catalunya, Barcelona, Spain

L. Blaškovičová
Slovak Hydrometeorological Institute, Bratislava, Slovak Republic

G. Blöschl & A. Viglione
Vienna University of Technology, Vienna, Austria

M. Borga
University of Padova, Padova, Italy

A. Dumitrescu, A. Irimescu & G. Stancalie
Meteorological Administration, Bucharest, Romania

S. Kohnova & J. Szolgay
Slovak University of Technology Bratislava, Bratislava, Slovak Republic

A. Koutroulis & I. Tsanis
Technical University of Crete, Chania, Greece

L. Marchi & E. Preciso
CNR, IRPI, Padova, Italy

Flash flooding can cause severe damage to buildings and infrastructure and pose a high risk to people yet by their nature, they are difficult to forecast. HYDRATE is a project funded by the European Commission, which is aiming to improve and develop techniques for flash flood forecasting. Work Package 1 (WP1) of the project is related to the collation of primary data on previous flash floods in the regions of the HYDRATE hydrometeorological observatories (see Table 1) and provides an essential resource to researchers.

The methodological approach for the collation of this primary data on past flash floods was to develop selection rules to establish which flood events would be included in the data set. The initial criteria for Phase 1 data collation were to record information on events that occurred between 1946 and 2007 on watersheds of an area of less than 500 km^2. The data set was then refined in Phase 2, selecting the most extreme events by identifying those events which are closest to the envelope curve for the region, i.e. that have high pseudo specific peak discharges. Data was collated using a standardised template containing sections on geographic, meteorological, hydrological and hydraulic data, as well as information on damages and casualties that were caused by the flood.

Table 1. Number of flash flood events listed in the HYDRATE WP1 database for each Hydrometeorological (HO) region.

HO Region	Number of events in Phase 1 selection	Number of events in Phase 2 selection
Catalunya, Spain	10	9
Cevennes-Vivarais, France	236	30
Northern Italy	73	30
Slovakia	52	30
Western Crete	21	4
Romania	152	30
Austria	34	17

The final data sets include 578 flood event records in seven European regions, which have been refined to 150 extreme events (see Table 1) for which full data templates have been completed where possible. Initial analysis of these data has been carried out to draw conclusions about the spatial and seasonal distribution of extreme events. The data sets provide the first step in the European Flash Flood database, which is being developed by the HYDRATE project. The work presented here represents some of the first outputs from this project.

Keywords: Flash flood, forecasting, data collation

Representative flash flood events in Romania Case studies

G. Stancalie, B. Antonescu, C. Oprea, A. Irimescu, S. Catana & A. Dumitrescu
National Meteorological Administration, Bucharest, Romania

M. Barbuc & S. Matreata
National Institute of Hydrology and Water Management, Bucharest, Romania

As in many European countries, floods are one of the most significant natural hazards in Romania, causing damage to buildings and infrastructure and posing a high risk to life. Among the floods events which took place in Romania in the last three decades many were flash floods.

HYDRATE is a project funded by the European Commission which main objective is to improve the scientific basis of flash flood forecasting, through analysis of past flash flood events at the European scale, advancing and harmonising across Europe a common flash flood observation strategy, developing and validating a coherent set of technologies and tools for effective early warning systems.

The paper presents a synthesis of the main characteristics of the Romanian representative flash floods events, included in HYDRATE Project.

Romania is characterized by a continental climate, with many periods of drought, followed by strong rainfalls produced in a short time. Heavy precipitation events in Romanian region typically occur downstream of a significant cyclone aloft, often exhibiting "cutoff" cyclone characteristics. Mediterranean and Black Sea proximity provide a source of moisture for Romanian territory and S-shape of Carpathians Mountains provide forcing. Large local rainfall totals often occurs when deep convective cells are organized such that they move repeatedly over a given area, a process commonly called "echo training". Size, organization and motion characteristics of the mesoscale convective systems are the principal factors that determine heavy rainfall. The convection can interact with its environment to develop new convection in preferred locations relative to the existing cells. When new cell move parallel to a outflow boundary (produced from the existing or previous cells), leaving a quasi-stationary segment of the boundary behind into which a substantial moist boundary-relative flow is impinging, creating new cells that repeat the motion of their predecessors. In addition to mesoscale convective systems, high precipitation super-cells have significant flash flood potential.

The ways by which atmospheric parameters for heavy precipitation are brought together can vary substantially from case to case and in different parts of the country. Heavy rainfall may or may not result in a flash flood, depending on the hydrological and geographical characteristics of the watershed where rainfall accumulates.

The ten severe flash flood cases analyzed in this paper are divided in three categories: 1) those resulting from intense rainfall in superior part of the natural watershed in the mountain regions, 2) those resulting from intense rainfall in basins altered by humans conductive of dam or levee failure and 3) those resulting from excessive rainfall in natural watershed in the plain regions. The analysis of these flash floods was made differentially, because some events took place in ungauged basins. In these cases, a post event survey was first realized and after that, through hydraulic methods the main hydrological parameters was determined.

The work is supported by the detailed presentation of three real case studies for which the most important meteorogical, climatological and hydrographical factors that lead to flash floods are emphasized.

Keywords: Flash flood, rainfall, convective systems

Changes in flooding pattern after dam construction in Zadorra river (Spain): The events of October 1953 and February 2003

A. Ibisate
Geography, Prehistory and Archaeology Department, University of Basque Country, Vitoria, Spain

Zadorra river is located in the headwaters of Ebro basin, it presents a basin of 1,361.28 km² and a mean discharge of 14.1 m³/s. In 1956 a reservoir system was constructed that presents three dams with a capacity of 147 Hm³, 72 Hm³ and 5.67 Hm³, respectively. This system has three main uses: energy generation, water supply and floods prevention.

Since Zadorra reservoirs came into operation Zadorra river's hydrology has notoriously changed. This has been also seen in the spatial location, intensity and danger of floods, and furthermore in the modification of the vulnerability of settlements and activities placed on the floodplain, and therefore the risk of these processes.

In this paper we present the modification of flooding pattern before and after the construction of the reservoir system. To illustrate and support the changes two different events are compared, one before the existence of reservoirs, October 1953, with a peak discharge of 364.23 m³/s and another one generated after that construction, in February 2003, with a peak discharge of 600.72 m³/s. Besides meteorological and hydrological data, it has been employed press and archives documentation to know the places flooded particularly in the October 1953 flood event.

One of the main conclusions of these two events comparison is that the spatial location of flooding area have already changed. Before reservoirs construction it was concentrated in the high stretch of the river, afterwards it was moved to the middle one, because of the water thrown by the dams and the "false sensation of security" generated by the reservoirs, which carried the more occupation of floodplain by settlements and industry.

Another conclusion is that the peak discharges have been increased, this is a direct result of reservoirs management and the difficult of harmonization of the different uses to which they are dedicated, two of them require the reservoirs to be as full as possible and the third one not so much, so that in many cases when the water is pour away to the river it must be done with high discharges.

So the conjunction of these two processes gives us a result of an increasing danger with a greater vulnerability that has lead to higher flood risk.

Keywords: Flooding pattern change, reservoir, risk, danger, vulnerability, Zadorra river

Post flash flood field investigations and analyses: Proposal of a methodology and illustrations of its application

E. Gaume
LCPC, Nantes, France

M. Borga
University of Padova, Italy

Flash-floods, rank as the most destructive process among weather-related hazards in many parts of the world. Foregoing studying these extreme events, because no measured data are directly on-hand or because they are not considered as sufficiently accurate, or even because it is time consuming, and limiting the hydrological analysis to moderate events on gauged catchments, would be focussing on the trivial while skipping the essential.

Post event survey and investigation is one way to gain experience on natural hazards. The importance of the systematisation and standardisation of such investigations and re-analysis is progressively recognised in all the geophysical sciences as shown by the growing number of scientific papers and programs on the subject. But how to proceed in the case of a flash floods, what type of data should be collected for what type of analyses and to explore which particular issues?

To give a first answer to these questions, a methodology for post flash flood field investigations has been developed under the EC FLOODsite project (Gaume, 2006). This methodology will be illustrated on two case studies: the analysis of the Gard (France) 2002 extraordinary flash floods and of the Selščica Sora, one of the catchments affected by the flash floods which occurred in September 2007 in Slovenia. 18 French hydrologists from 8 different institutions have participated to the Gard post flood investigation, while the Selščica Sora has been the first common European investigation involving researchers from 8 different countries and conducted in the context of the EC project HYDRATE. In both cases a large amount of cross sections were surveyed to map the peak discharges and identify the main contributing areas to the flood, witnesses were interviewed to gather information on the timing of the flood in the various areas of the considered catchments and rainfall-runoff simulations were done to connect the collected data and the radar rainfall estimates.

The analyses reveal a huge variability of the rainfall-runoff dynamics, depending on the characteristics of the affected catchments. Explanations based on the underlying hydrological processes supported by field observations can be proposed. In some cases, sub-surface flow seems to have played a dominant role even during very heavy rainfall events.

Gaume, 2006. Post Flash flood investigation, methodological note. EC FLOODsite project report D23.2. 62 pages.

Keywords: Flash floods, event analysis, rainfall-runoff

Hydrological and hydraulic analysis of the flash flood event on 25 October 2007 in North-Eastern part of Sicily, Italy

G.T. Aronica & G. Brigandì
Dipartimento di Ingegneria Civile, Università di Messina, Italy

C. Marletta & B. Manfrè
Dipartimento Regionale della Protezione Civile della Regione Sicilia, Servizio Provincia di Messina, Messina, Italy

On 25 October 2007, a deep cyclone developed in the Southern part of the Mediterranean basin produced an intense rainstorm over north-eastern part of Sicily, Italy. The storm event was produced by the clash of air sac of consistence and very different origins: to the ground, due to the break-in records of the last days, cold air sac still persist, and in altitude warm air sac transported by the sirocco flow above the persistent cold. The analysis of the event suggested that more than 120 mm of rain fell in less than 3 hours, although the 50% of the recorded rain fell in about 20 minutes.

The storm covered the southern part of the city of Messina and was concentrate on the Ionic sea coast near the village of Ali Terme in the catchment of Mastroguglielmo torrent.

That storm caused street flooding locally and, importantly, also overflows over portions of the lower reach of Mastroguglielmo river with significant damages to property, buildings, roads, bridges, and also several landslides that blocked the motorway and dragged the railway.

An issue that stirred the public opinion was whether the extensive overexploitation of the floodplain was responsible for the damaging overflows. We explore this issue by analysing the system with a rainfall-runoff model derive the flood hydrograph and with a two-dimensional flood propagation model to reconstruct the inundated areas. These model simulations support as plausible the hypothesis that flow obstructions due to the debris (reducing cross-sectional area of flow) caused the overflows.

Keywords: Flash flood, historical events, Inundation modelling

The day roads became rivers: A GIS-based assessment of flash floods in Worcester

F. Visser
University of Worcester, Worcester, UK

In the summer of 2006 the city of Worcester was hit twice by intense convective rainstorms, which caused severe flash floods outside of existing surface drainage networks. At the peak of the second event on July 6th, a total of 23.6 mm of rain fell within 35 minutes, which is half the average monthly rainfall. The Worcester fire brigade received 185 flood-related phone calls in the space of four hours. Worcester city council has collected 140 addresses from where flash flood damage was reported. The University of Worcester has compiled these in a GIS to enable analysis of processes that lead to the flooding in the affected locations.

The general public (incl. local newspaper, local authorities and affected residents) tends to ascribe the occurrence of the flash floods to the poorly maintained (Victorian) drainage system. It is thought to have too little capacity and is fully clogged up with sediment. In this paper this assumption is used to assess the occurrence of flow accumulation in Worcester in the absence of an artificial drainage system. The assessment also attempts to determine the relative importance of other factors that may contribute to the flooding (e.g. surface sealing, rainfall distribution). Only simple generally available tools and data will be used.

The Ordnance Survey DTM and the new OS MasterMap data form the topographic input for the assessment. MasterMap data is converted to raster format and used to create a preferential surface drainage system based on the roads network. Flow accumulation is calculated using the basic flow/hydrology modules of ArcGIS. Further analysis is done by dividing the whole of Worcester city into subcatchments. For each catchment an estimate is made of relative rainfall input based on radar data. For each catchment also a number of characteristics such as maximum flow accumulation and catchment area are recorded. A Landsat image is used to estimate imperviousness/roughness values, based on tasseled cap greenness transformation.

Flow accumulation estimates are compared with the spatial distribution of the flood damage records. Stepwise binary logistic regression is used to assess the importance of the various subcatchment characteristics for the recorded occurrences of flooding/flood damage. The results of the assessment are further used to discuss the validity of the claims made by the general audience on the drainage system and recommendations are made on how the results could offer an easy-to-use, pragmatic planning tool for policymakers who deal with flooding in Worcester.

Keywords: Flash floods, Flow accumulation

Risk and economic assessments

Flood risk mapping of Austrian railway lines

A. Schöbel
*Vienna University of Technology, Institute for Railway Engineering,
Traffic Economics and Ropeways, Vienna, Austria*

A.H. Thieken
alpS—Centre for Natural Hazard and Risk Management Ltd., Innsbruck, Austria

R. Merz
Vienna University of Technology, Institute for Hydraulic and Water Resources Engineering, Vienna, Austria

During the recent floods in Austria, the national infrastructure managers had to cope with important traffic disruptions on several line sections, involving—partly—also derailments and high amounts of loss. Since a great number of railway lines are located close to the course of rivers, traffic there consequently becomes strongly affected or impaired when flooding occurs. The normally inherent high safety and security level of rail traffic decreases considerably during and after flood events.

In this paper, a work flow towards a classification of railway sections on the basis of their damage potential is presented for Austria. The aim is to identify endangered and cost-intensive sections of the national railway network. For this, a national flood discharge map, containing inundation areas associated with 30, 100 and 200 year return periods for 26,000 km of Austrian streams, is combined with the assets of the railway infrastructure (bridges, tunnels, superstructures, signal boxes, etc.).

The flood discharge estimation is based on a novel concept with three guiding principles: 1) combination of automatic methods and manual assessments by hydrologists to allow speedy processing and account for the local hydrological situation; 2) combination of various sources of information including flood peak samples, rainfall data, runoff coefficients and historical flood data; and 3) involvement of the Hydrographic Services to increase the accuracy and enhance the acceptance of results. The estimated flood discharges have been transformed to flood hazards zones using hydraulic modelling.

The database of the Austrian railway infrastructure manager (ÖBB) covers a total of over 7,000 km of Austrian railways and contains information on all infrastructure-related issues including their asset values. For a first, country-wide flood risk assessment, the exposed asset values of all infrastructure elements, i.e. their damage potentials, were determined for the three flood hazard zones. From this, an expected annual damage potential was calculated for each element. In addition, the element's share in the total national expected annual damage potential was determined. With this approach, it becomes clear where valuable parts of the infrastructure are highly exposed to flooding and should be protected.

The combination of the flood hazards zones and the data base of railway infrastructure resulted in a flood risk map of Austrian railway lines, which is intended to be used by railway infrastructure managers for an optimized planning of flood protection measures.

Keywords: Flood risk analysis, railway infrastructure, exposure analysis, flood hazard zones

Correlation in time and space: Economic assessment of flood risk with the Risk Management Solutions (RMS) UK River Flood Model

D. Lohmann, S. Eppert, A. Hilberts, C. Honegger & A. Steward-Menteth
Risk Management Solutions Ltd., London, UK

Flood zone maps are an important instrument for the quantification of local flood risk, but are not useful in determining the risk of a portfolio (e.g. a whole country or an insurance portfolio of many houses). For that it is necessary to describe the spatial and temporal correlation of flooding. This can be done with scenarios that are subsequently used to analyse any arbitrary portfolio. The modelling system therefore has to be able to represent flood risk end-to-end for an entire country as a precipitation-runoff-flooding-loss-risk model up to return periods that are relevant to the insurance industry and decision makers, typically above 250 to 1000 years return periods.

Scenario models have been used for some time now in the insurance industry to quantify the risk of a portfolio. Often many peril models (hurricanes, winter storm, surge, hail, fluvial flooding) are integrated within a larger software platform that does the financial modelling. In 2001 RMS released the first version of the UK flood model. In 2008 RMS is releasing a major upgrade of this model in RiskLink Version 8.0. This presentation will discuss the basic methodology of risk models as well as key enhancements of the hazard model, the vulnerability model and the exposure distribution. The new UK flood model is based on a fine scale (10 m) DTM and the RMS European stochastic precipitation platform, and has been extensively verified with observed data from previous flood events. We will show new innovative solutions for on- and off flood plain probabilistic modelling that lead to an increased accuracy in calculating flood risk.

Keywords: Flood risk, probabilistic modelling, insurance

A case study of the Thames Gateway: Flood risk, planning policy and insurance loss potential

J. Eldridge & D.P. Horn
Birkbeck College, University of London, London, UK

In recent years there has been progress in managing flood risk with planning policy being strengthened to acknowledge flood risk at all stages of the planning process; however, developments continue to be built in high risk areas. This research aims to provide information for use by stakeholders involved with flood risk, and principally the insurance industry, to aid in the future development of insurance as a tool for flood mitigation. The study area is the Thames Gateway, which is currently undergoing a major regeneration scheme, the largest in Europe, creating considerable opportunity for housing and jobs. However, although this is beneficial to creating economic growth in previous industrial locations, many of the new developments will be situated on Thames floodplain land, resulting in potential exposure to flood events, in particular but not limited to an east coast storm surge. This poses the question of the level of risk that these properties face, a question particularly relevant after the events of the 2007 summer floods.

Of particular interest to insurance risk assessment is the loss potential to property in flood risk areas, with type of housing and density of most relevance. Linked to this is the effect of planning policy as a mitigation tool and how it contributes to risk under a range of scenarios. Currently, many different bodies, including governmental departments, private businesses and academia, are involved with issues in the Gateway; however, few attempts have been made to investigate the effect of such a development on insurance loss potential. This study will use modelling techniques designed specifically for insurance loss estimation to view future loss potential for existing and planned developments in order to characterise exposure and vulnerability within the built environment.

This paper presents the initial development of the project and the theory and methodology which will be used to address the research problem. The use of modelling techniques for inundation scenarios and catastrophe models for loss potential will improve understanding in prediction of the exposure that properties may face, producing a measure of flood risk for the Gateway area. Scenario based modelling of planning strategies for both commercial and residential properties will be used to identify the optimal approach in minimising flood risk. This study is being carried out in conjunction with the Willis Research Network, a body formed of academic collaborations overseen by Willis, a global insurance broker. The network contributes scientific understanding to an industrial setting and already has invaluable data available for input into this study.

Keywords: Flood risk, planning policy, insurance

Integration of accurate 2D inundation modelling, vector land use database and economic damage evaluation

J. Ernst
*Research Unit of Hydrology, Applied Hydrodynamics and Hydraulic Constructions (HACH),
Department ArGEnCo, University of Liege, Belgium*

B.J. Dewals
*Research Unit of Hydrology, Applied Hydrodynamics and Hydraulic Constructions (HACH), Department ArGEnCo,
University of Liege & Postdoctoral Researcher of the Fund for Scientific Research, F.R.S.-FNRS, Belgium*

P. Archambeau, S. Detrembleur, S. Erpicum & M. Pirotton
Research Unit HACH, Department ArGEnCo, University of Liege, Belgium

The Belgian national research project "*ADAPT—Towards an integrated decision tool for adaptation measures*", aims at developing a decision-support system (DSS) dedicated to the integrated evaluation of flood protection measures in the context of increased flooding hazard as a result of climate change. This DSS is based on a combination of cost-benefit analysis (CBA) and multi-criteria analysis (MCA) and takes into consideration hydraulic, economic, social as well as environmental parameters.

The present paper will focus on the integration between two components of the DSS, namely the evaluation of the economic impacts of floods and the hydrodynamic modelling. As described in a companion paper submitted for the conference (Dewals, B.J., S. Detrembleur, P. Archambeau, S. Erpicum, M. Pirotton: *Detailed 2D hydrodynamic simulations as an onset for evaluating socio-economic impacts of floods considering climate change*), the hydrodynamic simulations are conducted by means of 2D flow modelling and provide as an output high resolution flood maps detailing the distribution of water depth and flow velocity in the floodplains. The methodology will be illustrated for a case study along the River Ourthe in the Meuse Basin (Belgium).

The integration of hydrodynamic results and economic damage evaluation is based on four successive steps:

- combining inundation extent with a detailed *land use vector database*, enabling to identify the elements-at-risk (habitat buildings, industries, roads ...);
- based on the knowledge of the elements at risk, selection of proper *damage functions* and evaluation of the specific economic value of the elements-at-risk;
- combining flood maps (water depth and, possibly, flow velocities) with damage functions, leading to the relative damage (%);
- combining the relative damage with the *specific value* of the elements-at-risk, leading to the absolute damage (€).

The *land use database* includes 18 layers of information, among which following data is extracted and exploited: residence, industry, road network, agriculture (crops, fields) and forestry.

The *damage functions* used, linking the hydraulic parameters and the percentage of damage, can take various algebraic forms (continuous or not). Such functions are derived by relating water depth measurements and percentage of damage recorded for observed flooding events. The damage functions exploited here have been identified through a literature review (e.g. damage functions reported in the *ICPR Rhine atlas*).

The *specific values* of the elements-at-risk are gathered from several sources, such as local authorities or land registry offices ...

The paper will detail the application of the integrated assessment procedure for the case study of River Ourthe. Moreover, several examples of validation of the economic damage evaluation procedure will be provided by comparing computation results with real damage data recorded by the Belgian Disaster Fund after several major flood events (2003, 2002, 1995, 1993).

Keywords: Inundation modelling, economic impact of floods, damage function, land use, climate change, flood map

Planning for flood damages reduction: A case study

M. Karamouz
School of Civil Engineering, University of Tehran, Tehran, Iran

A. Moridi
School of Civil Engineering, Amirkabir University (Tehran Polytechnic), Tehran, Iran

A. Ahmadi
School of Civil engineering, University of Teharan, Tehran, Iran

In this paper, Kajoo watershed located in the south-eastern part of Iran that is affected by the summer monsoon and also winter floods is considered as a case study. A variety of structural and non-structural flood control options are used for reduction of flood damages in the study area. The non-structural flood control options consist of short term and long term activities in the study area for increasing system readiness to reduce flood damages. The short term activities are implemented considering a flood warning system which is based on precipitation and runoff volume at the specific gages during the flood in the study area.

The long term activities could be planned if a flood forecasting model could be implemented. In this study, the forecasting model (ANN model) is developed to process and train a system with large scale climate signals for the precipitation spells. In order to quantify the effects of large scale climate signals on the precipitation in the study area, long-term records of SST (Sea Surface Temperature) and SLP (Sea Level Pressure) over Oman Sea, Arabian Sea and Indian Ocean have been used. In this study, the HEC-HMS model is used for simulation of a flood hydrograph based on forecasted precipitation. In order to route floods along the river and floodplains, the HEC-RAS simulation model is used. Based on flood routing along the river, a variety of permanent and emergency flood control options are examined and structural flood control options for flood damage reduction are presented.

Keywords: Planning, Flood Damages, Precipitation Forecasting, Monsoon Rainfall, Flood Routing, Rainfall-Runoff Simulation

High resolution inundation modelling as part of a multi-hazard loss modelling tool

S. Reese
National Institute of Water and Atmospheric Research, Wellington, New Zealand

G. Smart
National Institute of Water and Atmospheric Research, Christchurch, New Zealand

New Zealand is well known for its scenery but little for its exposure to natural hazards. The extremes of weather and other natural forces that create its unique character also present many hazards, including earthquakes, volcanic eruptions, tsunamis, storms, floods and landslides. Flooding is by far the most common type of natural disaster in NZ and it has a major social and economic impact on affected communities. To plan for such catastrophies requires a clear understanding of the potential hazardous events, reliable hazard simulations and state of the art (or ark for floods?) risk analysis tools.

The paper describes how New Zealand copes with these challenges and presents recent research developments such as "RiskScape", New Zealand's first multi-hazard loss modelling tool. RiskScape is designed as a decision-support tool that converts existing hazard knowledge into likely consequences, such as damage and replacement costs, casualties, disruption and number of people that could be affected. RiskScape works by running through a sequence of steps which combine to simulate a risk profile for a region or locality. The cycle can be repeated for different hazards (e.g. tsunamis, earthquakes, floods & wind) simulating the way they could impact on the same cluster of communities and their associated assets. This allows a true comparison to be made of the potential consequences of different hazards on the same locality. Consequences for each region presented in a common platform can then form the basis of prudent planning and prioritized risk-mitigation measures that link directly to the severity of the risks.

A core part of RiskScape is accurate hazard modelling. In order to estimate for instance flood related damages, hydraulic parameters such as velocity and depth have to be associated with individual structures. This requires information at a very high resolution. Modelling is the only method of obtaining suitable data. In the past, computational constraints dictated the design and complexity of hydraulic model approaches. This has changed significantly with the advent of multi-processor, high speed desktop computers running two- or three-dimensional hydraulic models. Present computational models typically calculate water depths and velocities for around 5 million nodes at inter-distances as small as two metres. At this resolution, infrastructure such as roads and houses become part of a topographic elevation model. The modelling constraints have shifted from the hydraulic calculations to the provision of local ground heights and roughnesses at an accuracy and resolution commensurate with the computational model. The paper describes how remote sensing has been successfully applied to meet this requirement.

Flood modelling identifies the areas of risk of flooding, and provides the basis for estimating the consequences, severity, and cost to the region. The paper shows how sophisticated flood modelling, loss estimation and risk analysis can be combined in one single package and how this data can be used to improve the EM response, flood mitigation strategies and public awareness of flood hazard and thereby help local authorities with investment decisions for risk reduction for a region.

Keywords: High-resolution inundation modelling, loss modelling, RiskScape, multi-hazards, New Zealand research

Estimation of flood losses due to business interruption

I. Seifert, H. Kreibich & B. Merz
GeoForschungsZentrum Potsdam, Engineering Hydrology Section, Potsdam, Germany

A. Thieken
University of Innsbruck, Innsbruck, Austria

Increasing flood losses in the last decades have shown the urgent necessity to improve flood risk management. One of its crucial parts is the flood risk analysis, including loss assessment. For an exhaustive loss assessment all kinds of losses should be considered. However, research on flood losses and flood loss modelling, especially in the commercial and industrial sector, has not gained much attention so far. Up to now flood loss models mainly focus on direct losses to property. But particularly in the commercial and industrial sector, losses due to business interruption contribute a considerable part to the total losses. Therefore, in this contribution an approach for the consideration of flood losses due to business interruption is presented.

In the aftermath of the 2002, 2005 and 2006 floods in Germany extensive datasets about flood losses were collected via telephone surveys. Affected companies were questioned about their losses, precautionary and emergency measures, flood warning and lead time, flood parameters, etc. As far as possible answers were cross checked e.g. the absolute business interruption loss should be in the range of last year's turnover.

As Table 1 shows, water depth is the most important damage influencing factor concerning business interruption time. Other factors are precautionary measures, contamination and flood duration. The monetary loss caused by business interruption is described as business interruption loss per day. It was calculated as the quotient of absolute business interruption loss and business interruption time. Precautionary measures, company branch and company size (expressed as number of employees) have a significant influence on the monetary loss.

These findings were used to develop a model for the estimation of business interruption losses. The model works in two stages: First the mean business interruption time is deduced from information about water depth coming e.g. from simulated flood scenarios. In a second step the interruption time is linked to mean values for the interruption loss per day, which differentiate between company branches and sizes. As result we get an estimate of the absolute business interruption loss. If additional information on contamination and precautionary measures is available, scaling factors, which increase or decrease the business interruption loss, can be applied.

The applied model offers the possibility to estimate losses due to business interruption after flood events. The contribution shows the development, evaluation and application of the described model.

Keywords: Flood loss estimation, business interruption losses

Table 1. Factors influencing business interruption.

Factor	Business interruption time	Business interruption loss [€/day]
Water depth	$\rho = 0.35\ \alpha < 0.01$	no significant correlation
Flood duration	$\rho = 0.22\ \alpha < 0.01$	no significant correlation
Company branch	no significant correlation	$\rho = -0.29\ \alpha < 0.01$
Company size	no significant correlation	$\rho = 0.57\ \alpha < 0.01$
Precautionary measures	$\rho = -0.14\ \alpha < 0.01$	$\rho = 0.15\ \alpha < 0.05$
Contamination	$\rho = 0.22\ \alpha < 0.01$	no significant correlation

Correlation coefficient Spearman's ρ, level of significance α.

Residential flood losses in Perth, Western Australia

M.H. Middelmann
Geoscience Australia, Canberra, Australia

Perth, the capital city of Western Australia is situated on the Indian Ocean coastline. With a population exceeding 1.5 million, Perth is experiencing substantial population growth and is Australia's fourth most populous city. Increased demand for housing through migration and the resources boom has also increased house prices and there is an ever increasing pressure to develop land in potentially high flood risk areas. It is for these reasons that analysing the flood risk is vital.

The Swan River is the major river through Perth and flows into the Indian Ocean at the port of Fremantle. Flow was modelled using the one dimensional unsteady flow model HEC-RAS over a distance of 70 km. In total, seven streams were modelled including tributaries, over a combined distance of 188 km. The modelling of non urban areas was also considered significant because it identified flood prone areas where development may need to be restricted.

Direct economic losses to residential dwellings in Perth were modelled using geographical information systems as a tool for eight scenarios, ranging from the 10 year average recurrence interval (ARI) to the 2000 year ARI. Estimates of potential loss were modelled on an individual dwelling basis but were then aggregated. Estimates of structure costs were obtained using replacement cost models for Perth, indexed to 2008 values using Australian Bureau of Statistics data. Structure values were calculated from the structure cost multiplied by the floor area. Contents values were estimated as a percentage of structure which varied based on average household income per census collectors district and ranged from 20 to 40%.

Flood damage was assessed using synthetic stage damage curves developed for the two dominant wall types in Perth under both insured and uninsured scenarios. Data provided by a quantity surveyor on the repair costs of Australian structures were used in curve development. A separate contents curve was also developed, based on the estimated insured replacement cost. The contents curve assumed that no measures were taken to reduce the loss. This assumption was considered valid considering the lack of significant riverine flooding experienced in Perth during the preceding decades. Both structure and contents curves estimated damage as a percentage, which combined with data on structure and contents values, was used to estimate loss.

Combined structure and contents flood losses ranged from A$17 million to A$659 million for the insured structure and A$14 million to A$583 million for the uninsured structure. This equates to an average annual damage of A$9.6 million and A$7.9 million respectively. Understanding the spatial distribution of flood losses is important for targeting future flood risk mitigation strategies. Though 40 suburbs contribute to total flood losses, only 13 suburbs experience an insured loss exceeding a threshold of A$10 million. The suburbs of Ascot, Bassendean and Maylands experience the greatest flood losses, contributing a combined total loss of 41% during the 2000 year ARI scenario. However, residents in the suburbs of Ascot, Viveash and Guildford are most likely to be flood affected, based on the relative proportion of dwellings in each suburb that are flood affected. The results reinforce the need to consider a wide range of varying magnitude flood events when assessing loss because of the temporal and spatial variation between scenarios.

Keywords: Average annual damage, geographical information systems, HEC-RAS, residential flood losses, spatial variation, stage damage, temporal variation

A multicriteria flood risk assessment and mapping approach

V. Meyer & D. Haase
UFZ—Helmholtz Centre for Environmental Research, Leipzig, Germany

S. Scheuer
Martin-Luther-University Halle-Wittenberg, Institute for Geosciences, Halle, Germany

Flood risk analysis and assessment are integral parts of the flood risk management approach. However, some deficits can be recognised in today's practice with regard to the following aspects:

a) The focus of flood risk assessment is still very much on economic flood risks. Social and environmental flood risks are often neglected. Consequently, the results of risk assessment can be incomplete and biased. b) The spatial distribution of risks as well as of the benefits of flood mitigation measures is rarely considered. E.g. the evaluation of mitigation measures is mostly based on their overall effect. Therefore, it is often not considered which areas benefit most from a measure and which areas do not. c) Uncertainties in the results of risk assessment are often ignored.

In this paper we want to present a GIS-based multicriteria flood risk assessment and mapping approach. Our approach can be used for an integrated assessment of economic, as well as social and environmental flood risks. Furthermore, the spatial distribution of these multiple risks as well as of the effects of risk reduction measures can be shown by this mapping technique. Moreover, possibilities are shown how to deal with uncertainties in criteria values and to demonstrate their influence on the overall assessment.

The approach is applied to a pilot study at the River Mulde in Saxony, Germany, heavily affected through the hazardous flood in 2002. Therefore, a GIS-dataset of economic as well as social and environmental risk criteria was built up, using the criteria shown in table 1. Two different multicriteria decision rules, a disjunctive approach and an additive weighting approach are used to come to an overall assessment and mapping of flood risk in the area. Both, the risk calculation and mapping of single criteria as well as the multicriteria analysis are supported by a software tool (FloodCalc) which has been developed for this purpose.

Keywords: Flood risk assessment, flood damage evaluation, multicriteria analysis, risk mappi Young FLOODsite Prize Competition!

Table 1. Criteria used for the multicriteria risk assessment approach.

Risk Dimension	Criteria
Economic	Annual average damage
Social	Annual average affected population
	Probability of social hot spots (hospitals, schools etc.) being affected
Environmental	Erosion potential (of material)
	Accumulation potential (of material)
	Inundation of oligotrophic biotopes

New developments in maximizing flood warning response and benefit strategies

S.J. Priest, D.J. Parker & S. Tapsell
Flood Hazard Research Centre, Middlesex University, London, UK

Flood forecasting and warning systems have a significant role to play within integrated flood risk management, either in combination with mobile structural flood defences or as part of an approach which combines a number of non-structural measures. In theory the benefits of flood warnings, in terms of community security, protection of life and flood damage reduction, should be large. The theoretical benefit potential is being driven upwards by important advances in the predictive sciences, and in the application of information and communication technologies to rapid flood warning communication. However, in practice the full benefit potential of flood warnings is rarely maximised, and currently in some circumstances it falls well below the theoretical potential. Often this happens because warning response strategies are inadequately conceptualised, developed and implemented. The position has not been helped by data being unavailable on the loss saving potential of different warning response strategies.

This research from FLOODsite Task 10 reconceptualises flood warning response strategies within a portfolio approach context, and develops a comprehensive model of flood warning response which may be applied to a wide range of flooding conditions and socio-economic circumstances. The model allows different flood warning response strategies to be configured and designed for the unique circumstances of each flood location, usually in combination with other measures.

Two models are now available for estimating the benefits of flood warnings in terms of flood loss savings—for the first time providing data on the loss saving potential of each strategy which can be fed into decisions designed to maximise these flood warning response and benefit strategies. One model focuses in detail on household flood warning response and benefit, while the other takes a much broader approach considering benefits which include those generated to all sectors of an economy by deploying mobile flood defences, through to benefits generated by business continuity planning. Quantitative results from a) a synthetic simulation of flood warning response and benefit in three urban areas, and b) an application to a recently flooded urban area on continental Europe are presented. They demonstrate the potential of the reconceptualised flood warning response strategies, and the comparative scale of monetary benefits associated with each. The factors which have been found in the research to encourage improved flood warning response, and thus warning benefits, are also presented and discussed.

The models, data and findings from this research are now ready to be applied throughout Europe, and are capable of significantly increasing the value of flood forecasting and warning systems. Despite progress some significant data gaps still remain and provide an agenda for future research in this area.

Keywords: Flood warnings, warning response, warning benefits, Europe

Development of a damage and casualties tool for river floods in northern Thailand

J.K. Leenders, J. Wagemaker & A. Roelevink
HKV-consultants, Lelystad, The Netherlands

T.H.M. Rientjes & G. Parodi
International Institute for Geo-Information Science and Earth Observation—ITC, Enschede, The Netherlands

For many people around the World and particularly in developing countries the dangers associated with river floodings are serious. Houses can be destroyed and land used for agricultural purposes can be affected. The impact of flooding on society can be dramatic and in recent months floodings in China and Bangladesh caused thousands of people to become homeless or displaced. The discussions on climatic change and sea-level rise created awareness and questions are raised on what to do to prevent or mitigate the negative impacts of flooding and how to adapt to it.

Understanding the effects of flooding in terms of effects on economic losses and casualties plays a key-role in this discussion. Such understanding facilitates decision-making on regional planning and flood control measures by presenting its benefits in terms of economic value. Research on damage assessment has mostly been carried out in the USA, UK, Australia and the Netherlands. However, these countries commonly are not frequently and also not seriously affected by river floodings. Countries as Thailand, Bangladesh, Indonesia and India and others in the Pacific typhoon belt, however, report on serious floodings almost every year. Flood damage assessment in these countries is difficult since damage assessment tools as developed are not directly applicable. While land-uses and commercial activities are different as compared to western countries also the damage functions themselves differ.

In this study a tool is developed to assess damage and casualties as a result of flooding in a study area in Thailand. The tool is GIS-based and gives insight in economical losses and losses of casualties as a result of potential floods. The paper discusses the following:

1. Assessment of infrastructure and demographic data of the study area using remote sensing.
2. Development of damage functions that relate effects of flooding to the infrastructure and to demographic data.
3. Assessment of damage values related to different commercial activities and land-use types.

By combining these aspects and results of a 1D2D hydraulic flood model an example is presented on effectiveness of flood damage assessment in Thailand. In this study also the potential economic value of the area is assessed. It is shown how the damage assessment tool can be used in decision-making on regional planning and design of flood control measures using the present and potential economic value of the area.

Keywords: Damage assessment, 1D2D flood modelling, remote sensing

Synthetic water level building damage relationships for GIS-supported flood vulnerability modeling of residential properties

M. Neubert, T. Naumann & C. Deilmann
Leibniz Institute of Ecological and Regional Development (IOER), Dresden, Germany

Climate change will lead to a further increase of elements and values at risk. The European Directive on Flood Risk Management accordingly will require estimations of vulnerability and flood risks in the long term. Therefore, tools are needed which are able to simulate scenarios of future developments. The paper will focus on flood damages to residential properties. The presented Damage Simulation Tool was developed within the research project VERIS-Elbe funded by the German Ministry for Education and Research (BMBF). The strength of the approach is the high spatial resolution of the model which combines building mask GIS and synthetic damage functions (refurbishment costs) along building types.

Up to now vulnerability assessment in terms of economic damage potentials was merely based on empirical damage functions from the last decades and expert knowledge. Although a lot of socioeconomic damage potential analyses have already been done, most of them have been based on status quo land use patterns and rarely considering significant land use changes and increasing economic values. In the perspective of a time horizon of 50 years, especially for residential building structures, completely new damage functions are needed based less on statistics but on architectural and engineering know how on cause-effect relationships and refurbishment costs. Furthermore: It might be possible for small investigation areas to do a house by house on site investigation, but when it comes to analyses and modelling of large river basins this is impossible. Then the next resolution level will have to handle with a typology approach for the building stock.

The simulation model (HOWAD—Hochwasser-Schadens-Simulations-Modell) for potential damages of buildings and constructed assets is based on a new bottom up approach. At the core stands the Urban-Structural-Type (UST) approach. UST's are areas with physiognomic homogeneous characters of built up areas marked by characteristic formation of buildings and open spaces. The UST can be linked to a building typology, where representatives of buildings by age and type can be allocated. Special descriptions of each representative include construction details, which are relevant for refurbishment. By that, modes of construction, quantity structure, materials and typical damages due to floods and refurbishment cost can be assigned. A highly differentiated GIS-supported analysis of built up areas connects the data and merges the UST specific database with spatial distribution data and occurrence of hazards. Thus, single buildings detected by the Digital Elevation model in the flood plain are recognized by their UST-specific attributes—using ArcGIS. The model combines the water level with the building type and its synthetic damage functions in order to calculate the vulnerability. The data can be aggregated for regions of interest (administrative districts, catchment areas etc.).

The strength of this new approach is a highly detailed model especially for the housing areas damage potential based on refurbishment costs. For industrial areas a real estate average value approach will be applied together with a scoping in order to detect hot spots which need further detailed enquiries. The combination of both object-based approaches allows a highly differentiated evaluation of the impacts of changing risks by climate and societal change. Moreover, it allows ex ante analyses of mitigation strategies.

Keywords: Flood risk management, damage functions, vulnerability modelling, geographic information systems

Impacts of the summer 2007 floods on agriculture in England

H. Posthumus, J. Morris, T.M. Hess, P. Trawick, D. Neville, E. Phillips & M. Wysoki
Cranfield University, UK

Exceptional rainfall during the summer months in 2007 caused extensive flooding in parts of England and Wales, resulting in the displacement of people and extensive damage to houses, businesses and infrastructure. While the focus of attention has been correctly placed on the impact on life and property in densely populated urban areas, large tracts of rural land were seriously affected by flooding. In many cases, rural land, especially that occupied by agriculture, is naturally or purposefully used to temporarily store floodwaters. The timing of the 2007 flood events coincided with the period when rural land is particularly 'productive', that is, during harvest period for agriculture. Furthermore, the summer floods occurred during the busiest period for visitors enjoying the countryside. This makes rural areas, and the people and businesses that occupy them, particularly vulnerable to summer floods.

Changes in rainfall intensity and distribution associated with climate change are expected to increase the extent and frequency of flooding in future. Flood risk, defined as the probability of a flood event multiplied by its economic cost, is highest in urban areas. Protecting the relatively 'sparse' rural areas against flooding is therefore increasingly deemed to be insufficiently cost-beneficial compared to urban areas. Under the UK flood risk management strategy Making Space for Water, rural floodplains are increasingly designated to store floodwater to protect downstream urban areas from river flooding.

This paper presents the results of a study to evaluate the immediate and potentially long term consequences of the summer 2007 flood events for rural households, businesses and communities. Surveys were carried out of households, businesses and farmers in flooded rural areas in three regions in England, namely: Yorkshire, Worcestershire and Gloucestershire, and Oxfordshire. The results seek to determine whether there is a particular 'rural dimension' to the extent and impacts of, and responses to, flooding of the kind experienced in summer 2007, in ways that distinguishes these from the experiences in urban areas. Understanding the consequences of flooding in rural areas and the associated costs for society can help to ensure that the particular needs and vulnerabilities of rural areas are explicitly considered in the formulation of future strategies to manage flood risk.

Keywords: Flood risk management, summer 2007 floods, rural society and economy, England

Climate change

Simulating flood-peak probability in the Rhine basin and the effect of climate change

A.H. te Linde
Institute for Environmental Studies (IVM), Vrije Universiteit, Amsterdam, The Netherlands
Deltares, Delft, The Netherlands

J.C.J.H. Aerts
Institute for Environmental Studies (IVM), Vrije Universiteit, Amsterdam, The Netherlands

Flood control and water resources management activities in the Rhine basin started already in the beginning of the 19th century. Since the flood events of 1993 and 1995, and more recently the growing awareness that climate change does significantly effect weather patterns and therefore the runoff regime, flood risk management has gained attention in the region, resulting for example in the EU Flood Directive.

Models, datasets and understanding of the hydrological system continue to improve and scientific studies on the effects of land use change, river training measures and climate change on the discharge regime increasingly feed flood policy discussions. Few studies, though, have studied the combined effect of these variables in scenarios exploring possible future changes. Furthermore, both scientists and policy makers recognize the inherent uncertainty when performing future scenario studies, which resulted in the wish to quantify and, if possible, reduce this uncertainty.

To create forcing data for the hydrological model (HBV) under climate change, the regional climate model (RCM) RACMO was used. Biases in precipitation and temperature were accounted for by correction factors. RACMO output was used both directly as input to HBV, and to define basin wide mean changes to transform historical rainfall and temperature data by the delta-method. The safety levels in the downstream part of the Rhine basin are high: up to return periods of 500 years in the Lower-Rhine area in Germany and 1,250 years in the Netherlands. This results in high uncertainties when fitting extreme value distributions to historical or calculated discharges, which are in the order of decades. In an attempt to reduce this uncertainty, a re-sampling strategy was applied on the forcing data, resulting in input series of 1,000 and even 10,000 years. Using these data, 1,000 years of daily discharges were calculated by HBV. Yearly maximum peak flows were then re-calculated using the hydrodynamic model SOBEK, which describes the main Rhine branch and parts of several side branches, and allows for implementing structures and measures like detention areas, river widening or dike heightening. Sets of measures are combined in adaptation strategies and implemented in SOBEK. Results were analysed at several locations throughout the Rhine basin.

It can be seen that the effect of climate change is expected to be considerate, both on mean discharges and on extreme events. Winter discharges are expected to increase and summer discharge to decrease. Discharges and probabilities of extreme flood events are expected to increase. Effects of measures vary due to differences in flood height, flood wave generation and, in case of controlled detention areas, timing.

The current paper displays a method to determine the relative contribution of both climate change and adaptation strategies on flood generation in the Rhine basin. It is argued here that studying future scenarios using re-analysis time series up to 1,000 years of daily values, provides insight in the heterogeneity and uncertainty of weather patterns, peak flow events and effects of adaptation measures. This will add to the knowledge on flood management strategies under climate change, especially when effects of measures on extreme events are considered.

Keywords: Flood risk, extreme events, climate change, adaptation strategies, HBV, SOBEK

Climate changes in extreme precipitation events in the Elbe catchment of Saxony

C. Görner, J. Franke & C. Bernhofer
Technische Universität Dresden, Institute of Hydrology and Meteorology, Dresden, Germany

O. Hellmuth
Leibniz Institute for Tropospheric Research, Leipzig, Germany

Flooding—including flash floods—is the most widely distributed of all natural hazards across Europe, causing distress and damage wherever it happens. To avoid the flooding of populated areas flood protection concepts are necessary which have to be well adapted to the actual climate conditions and its changes in the future, especially concerning heavy precipitation events. Therefore, regionalised climate change scenarios are required.

Within the work of the meteorology team of the TU Dresden concerning the Pilot Study "River Elbe Basin" (Task 21) two climate scenarios were analysed for Saxony and the Upper Mulde River Basin. These scenarios have been calculated by *ENKE* based on the IPCC climate scenarios A2 and B1. For the data analysis a weather generator for temporal downscaling was developed. The structure and functionality of this weather generator will be presented schematically.

Furthermore, the projected climate changes for Saxony will be presented exemplarily. The focus is thereby on precipitation. It has been examined which changes in the amount, frequency, intensity and return period are projected, especially of heavy rainfall events. Concerning the return period of extreme precipitation events today's 100 yr. precipitation was chosen with durations of 24, 48 and 72 hours. The respective return period under conditions of regional climate change was calculated by the KOSTRA method, as used in Germany for the calculation for design precipitation. Statistical downscaling was applied to generate the necessary extreme value distribution parameters.

With the assumption of a runoff coefficient close to one the occurrence of heavy rainfall will be similar to the occurrence of heavy floods. Therefore, based on these analyses a first guess on the occurrence of future floods and there damage potential is possible.

Keywords: Flood risk management, climate change, extreme precipitation, return period

A methodology for adapting local drainage to climate change

R.M. Ashley & J.R. Blanksby
Pennine Water Group, The University of Sheffield, England, UK

A. Cashman
University of the West Indies, Barbados

R. Newman
Pennine Water Group, The University of Sheffield, England, UK

This paper is based on the recently completed AUDACIOUS project which has developed adaptive responses for coping with climate change for local drainage systems based on both new computational models and also frameworks for responding. A methodology is presented that will allow all stakeholders to play their appropriate part, especially as everyone is at risk of flooding from climate changed rainfall patterns in local urban areas. In the UK, communication of this risk has so far been very poor and left to distorted media reporting and blame passing by key stakeholders[1]. At a property level, there is a lack of information, engagement and connection by individuals to the consequences of climate change on property management. The paper illustrates through 6 case studies, how the complex nature of flow interactions, even at the local scale, necessitates a more holistic, flexible approach to the design, operation and management of building and local drainage systems. It is clear that individual drainage system elements may not operate as expected under the type of extreme loading expected under climate change. The overall approach is one in which adaptive capacity should be developed through the cultivation of whole-system resilience (not the commonly used narrow definition applying only to properties), to include building capacity of both stakeholder groups and also in local drainage systems. This will require the development of 'active learning' cultures across the population. Sadly in England, current institutional attitudes and structures militate against effective engagement, active learning and capacity building by overly complex institutional arrangements and cross-institutional barriers in the water and flood risk management arena.

Overall this creates confusion in the dwellers, property owners and occupiers. Exhortations to this group to become more involved in responding are futile without more and effective capacity building. Flood management policies devised at national level are often too remote from local needs. To be effective there needs to be much more management at a local scale with devolved budgets and responsibilities. This will become increasingly more important as local areas identify their own increasing flood risks and seek to take action. The insurance industry also has a key and valuable role in ensuring that Government have the right institutional arrangements in place to make flood risk management sustainable. The insurance industry should make more effort to positively contribute to developing responsive capacity as well. The current system of valuing and 'surveying' properties for mortgage purposes in the UK fails to account adequately for flood risk highlighting the need for more specialist professional water management input. The current information required in the Home Information Packs (HIPs) fails to account for the growing need to ensure that flood risks are adequately highlighted when properties are purchased[2].

The transfer of communal private sewers to the English and Welsh sewerage undertakers in 2010 will be beneficial in ensuring a level of service but this will also confuse and further distance local property dwellers from the need to take an active interest in their drainage systems. The undertakers may also find it difficult to deliver the expected levels of service especially as the condition of the transferred sewers is largely unknown. Experience with the water industry regulator Ofwat also suggests that the sewerage undertakers will not be

---

[1] See for example UK EA report on summer 2007 flooding.
[2] This is highlighted in the recent Pitt inquiry interim report.

adequately funded to provide the expected service. There is also a need to review the way in which regulations and standards are defined for buildings and associated drainage. There needs to be a move towards standards based on 'quality of life' or 'health and welfare' requirements and should be considered in the current review of building regulations taking place in England and Wales.

Keywords: Flood risk management, adapting to climate change, capacity building, case study

Exploring and evaluating futures of riverine flood risk systems – the example of the Elbe River

J. Luther & J. Schanze
Leibniz Institute of Ecological and Regional Development (IOER), Dresden, Germany

Flood risks occur in human-environment systems and underlay a long-term dynamic caused by a number of drivers such as climate change, land-use changes, and others. Many drivers evolve slowly over time or show time-lag effects and long return periods. Moreover, certain decisions may determine the control actions of the following decades. This calls for a long-term view into the future and implies the challenge of dealing with many uncertainties due to the system's complexity. The paper presents a planning approach for exploring, analysing, and assessing 'alternative futures' of riverine flood risk systems with the example of the Elbe River basin. Futures in this respect are understood as consistent assumptions on the change of system factors considering autonomous developments (scenarios) and potentials for control (strategic alternatives). It is argued that this approach is a way of identifying the impacts of possible developments and actions that can reveal the scope of uncertainties inherent to the flood risk system's dynamic and the performance of risk reduction options.

- A first step is the conceptualisation of a holistic flood risk system following the SPRC-model. Its physical geographical and anthropogenic factors may either be subject to autonomous trends, target-oriented control, or facultative system behaviour (e.g. dike breaches). With this concept, the integration of different processes and scales is aspired. The comprehensive description and simulation of riverine flood risk systems with a consistent exploration of its factor dynamics on both the hazard and the vulnerability side are still in its infancy in flood risk management research.
- Secondly, it is shown how the 'risk cascade' for present and future states of the flood risk system can be calculated based on coupled models ranging from climate change projections to a high resolved damage simulation model. In this step it becomes clear to what extent factors and their dynamics can be included and which are constraints due to data and model availability, calculation time, personal resources, and so forth.
- Thirdly, socioeconomic storylines for the scenarios and guiding principles for the strategic alternatives are developed and the futures are combined. This involves making plausible and consistent assumptions for many system factors and their drivers and finding ways to harmonise existing data for the same areas and time steps. Both global and regional aspects are considered.
- Fourthly, selected futures are analysed ex ante applying the coupled models to derive the emerging flood risks.
- As a last step the evaluation addresses, amongst other aspects, the identification of (i) the sensitivity of *all* scenarios against the current strategic alternative; (ii) the resulting risks when applying different strategic alternatives against *one* selected scenario; (iii) the efficiency (as cost-effectiveness) and robustness of *one* selected strategic alternative against the different scenarios; and (iv) the model uncertainty, for example caused by different climate downscaling methods.

The conceptualisation of the flood risk system, the coupled modelling, and the scenario design have been carried out under the German national research project 'Change and Management of Extreme Flood Events in Large River Basins—the Example of the Elbe River' (VERIS-Elbe) which is funded by the Federal Ministry of Education and Research (BMBF) under the National Research Programme 'Risk Management of Extreme Flood Events' (RIMAX), whereas the evaluation methodology is an outcome of FLOODsite Task 14.

Keywords: Flood risk management, scenario planning approach, long-term planning, Elbe River basin

Note: Please consider the abstract/paper for the Young FLOODsite prize.

Author index

Acosta-Michlik, L. 79
Adams, B. 186
Aerts, J.C.J.H. 61, 299
Ahmadi, A. 287
Allitt, R. 19
Allsop, N.W.H. 99
Allsop, W. 89, 93, 98, 122
Alvarado-Aguilar, D. 70
Andrés, A. 225
Andrieu, H. 210
Anquetin, S. 174
Antonescu, B. 275
Archambeau, P. 31, 286
Arnaud, A. 190
Aronica, G.T. 22, 226, 278
Arránz-Becker, O. 146
Ashley, R. 132, 157
Ashley, R.M. 301
Asselman, N. 26, 47
Assmann, A. 46, 48, 196
Astle, G. 220
Asztalos, J. 211

Bachmann, D. 119, 120, 146
Bain, V. 273
Bakonyi, P. 148
Ball, L.G.A. 144
Ball, T. 81, 128
Bally, Ph. 198
Balmforth, D.J. 67
Bamford, T. 67
Barbuc, M. 273, 275
Barnes, A. 217
Barr, S.L. 114, 128
Bateman, A. 273
Bates, P. 47
Bates, P.D. 26, 54
Batty, L. 249
Batty, M. 114
Bazin, P.-H. 93
Bell, G. 144
Beneš, V. 121
Bernardara, P. 190, 273
Berne, A. 187
Bernhofer, C. 219, 262, 300
Beven, K. 56, 216, 221

Black, A.R. 81
Bladé, E. 16
Blair, G. 216
Blanksby, J.R. 301
Blaškovičová, L. 273
Bless, J. 146
Bliefernicht, J. 196
Bloomfield, J.P. 186
Blöschl, G. 130, 273
Blum, H. 234
Blume, T. 130
Bogacz, P. 117
Boonya-aroonnet, S. 19
Borga, M. 148, 187, 219, 224, 273, 277
Borthwick, A. 52
Boudevillain, B. 187
Boukalová, Z. 121
Boukhris, O.F. 189
Bouwer, L.M. 61
Bramley, M. 112, 116
Braun, L. 211
Bray, M.T.J. 212
Brigandì, G. 226, 278
Bronstert, A. 130, 241
Bruce, T. 89, 99
Bubeck, P. 61
Buber, A.L. 200
Buijs, F. 109
Buijs, F.A. 91, 122
Burg, S. 234
Butler, J.B. 21
Butts, M.B. 66

Cai, Y. 183
Cali, M. 249
Calver, A. 181
Candela, A. 22
Cashman, A. 157, 301
Catana, S. 275
Catovsky, S. 86
Cavalli, M. 66
Chan, A.H. 73
Chapon, B. 187
Chatterton, J. 264
Chavet, I. 93

Chemitte, J. 257
Chen, Y. 33
Chenaf, M. 45
Chini, N. 79
Clausen, T. 66
Clegg, M.J. 144
Cluckie, I. 33, 212
Cluckie, I.D. 82
Cobby, D. 220
Colbourne, L. 153
Colin, O. 198
Coller, M.L.F. 43
Collier, C.G. 77
Cooper, A. 17
Corestein, G. 16
Corral, C. 134, 223
Cossu, R. 198
Coulet, C. 18
Coulson, G. 216
Courage, W.M.G. 65
Craciunescu, V. 50
Creutin, J.-D. 174, 224
Creutin, J-D. 78
Crossley, A. 27
Crossman, M. 260
Cuomo, G. 41
Cuppen, M. 161
Cuppen, M.E. 172
Czigány, S. 245

Daamen, K. 196
Dagoumas, A. 114
Dale, A.G.J. 263
Dale, M. 217
David, E. 199
Dawson, Q.L. 151
Dawson, R. 68, 79
Dawson, R.J. 114, 128
de Boer, E. 109
de Bruijn, K.M. 142, 172
De Marchi, B. 167, 204
de Raat, G. 95
de Rocquigny, E. 190
de Vries, W. 95
de Wit, A. 26, 47
Deeming, H. 165, 202

Deilmann, C. 294
Delobbe, L. 59
Delrieu, G. 187
Denness, D. 116
Dent, J. 220
Deppe, T. 86
Deroubaix, J.F. 132
Detrembleur, S. 31, 286
Dewals, B.J. 31, 286
Di Mauro, M. 195
Dickson, M. 68
Digman, C.J. 67
Dister, E. 131
Dolz, J. 16
Doncaster, S.H. 154
Doorn, N. 122
Dorner, W. 127
Douglas, I. 132
Drabek, U. 218
Dufour, F. 209
Duggan, S. 249
Dumitrescu, A. 273, 275
Dunning, C.M. 168
Dunning, P. 185
Durden, S. 168

Ebert, C. 196
El Tabach, E. 132
Eldridge, J. 285
Enggrob, H.G. 30
Eppert, S. 284
Erlich, M. 199
Ermolaeva, O.S. 200
Ernst, J. 286
Erpicum, S. 31, 286
Escher-Vetter, H. 211
Ettrich, N. 36
Evaux, L. 18

Fabio, P. 22
Falconer, R. 220
Falconer, R.A. 253
Feger, K.H. 250
Fernandez-Bilbao, A. 153, 202
Fewtrell, T. 47
Fewtrell, T.J. 26, 54
Fielding, J.L. 169
Filatova, T. 259
Finlinson, B. 171
Flikweert, J-J. 112
Flowerdew, J. 231, 232
Flueraru, C. 50
Fontaine, C. 79
Fontenot, E. 66
Forbes, G. 220
Forcadell, D. 223
Ford, A.C. 114, 128

Förster, S. 241
Fortune, D. 44, 141
Francés, F. 130
Francis, O. 40
Francke, T. 130
Franke, J. 300
Frenzel, H. 250
Fries, J. 146
Frissel, J.Y. 95
Frogbrook, Z. 40
Fuchs, S. 127
Funke, R. 213

Gaál, L. 180
Galiatsatou, P. 182
García, E. 225
Garcia, J. 273
Garcia, R. 41
Garvin, S. 132
Gaume, E. 210, 222, 224, 273, 277
Geisenhainer, P. 105
Gerber, M. 209
Geresdi, I. 245
Gillon, S. 157
Gleitsmann, C. 83
Gocht, M. 134, 136
Golding, B. 220
Gómez-Valentín, M. 16
Gonzalez-Marco, D. 177
Görner, C. 219, 300
Goudie, J. 171
Gouldby, B. 15, 17, 109, 112, 148, 269
Gouldby, B.P. 101
Goutal, N. 190
Goutiere, L. 47
Gowing, D.J. 151
Grimaldi, S. 41
Gruntfest, E. 174
Gupta, S. 109
Gutierrez Andres, J. 17

Haase, D. 291
Habersack, H. 131
Hall, J.W. 68, 79, 91, 114, 128
Han, D. 33, 212
Hannan, J.M. 55
Hanson, H. 236
Hanson, S.E. 79
Hardy, K. 264
Harms, O. 131
Harries, T. 171, 173
Harteveld, C. 113
Hartnack, J.N. 30
Hassan, M.A.A.M. 105
Hauer, C. 131

Hawkes, P. 98
Hawkes, P.J. 232
Hazenberg, P. 59
Hecker, E.J. 170
Hedges, T.S. 235
Hegg, C. 209
Hellmuth, O. 300
Henderson, D. 150
Hendy, P. 162
Hennersdorf, J. 196
Hess, T.M. 151, 295
Hilberts, A. 284
Hilker, N. 209
Hlavčová, K. 180
Hofmann, S.D. 69
Hollá, M. 215
Holubova, K. 156
Holzhauer, V. 46
Honegger, C. 284
Hope, I.M. 193
Höppner, E. 48
Horn, D.P. 285
Horrillo-Caraballo, J.M. 246
Horritt, M.S. 54
Horsburgh, K. 232
Horsburgh, K.J. 231
Hounjet, M. 113
Houston, D.M. 81
Hu, K. 235
Huber, N.P. 107, 120, 146
Hughes, A.G. 186
Hughes, A.K. 193
Hughes, D. 216
Hunt, A. 52
Hunter, N.M. 54, 67
Huthnance, J.M. 242
Hutter, G. 138, 155

Ibisate, A. 276
Irimescu, A. 50, 273, 275

Jackson, B.M. 40
Jackson, V. 86
Jatho, N. 219
Jensen, T.S. 66
Jiménez, J.A. 70, 148
Jobe, M. 81
Johnson, C. 258
Jones, D.A. 179
Jonoski, A. 49
Jude, S.R. 79

Kaczmarek, J. 117
Kahl, B. 134, 227
Kaiser, G. 69
Kandasamy, J. 55
Kanning, W. 109

Karamouz, M. 287
Karunarathna, H. 242
Kashefi, E. 153, 163, 164
Kasheri, E. 202
Keef, C. 185
Khatibi, R. 74
Kingston, G. 109, 269
Kingston, G.B. 101
Kirnbauer, R. 211, 218
Kirstetter, P.-E. 187
Kjeldsen, T.R. 179
Klijn, F. 78, 142
Kohnova, S. 273
Kohnová, S. 180
Komma, J. 130
Köngeter, J. 107, 146
Königer, P. 138
Koppe, B. 71
Kortenhaus, A. 69, 78, 99, 105, 109, 122
Koutroulis, A. 273
Krahe, P. 196
Krämer, S. 72
Kreibich, H. 289
Krischke, M. 48
Kron, A. 36
Krywkow, J. 172, 259
Kufeld, M. 119
Kuhlicke, C. 167
Kunapo, J. 43
Kunz, M. 196
Kyselová, D. 215

Lamb, R. 27, 185
Lambrecht, H.-J. 109
Lane, A. 235, 242
Larson, M. 236
Lawless, M.R. 38
Lawson, N. 132
Leake, J. 79
Leedal, D. 56, 221
Leenders, J.K. 293
Leijnse, H. 59
Leitão, J.P. 19
Lennartz, G. 146
Lešková, D. 215
Lesniewska, D. 117, 122
Lewis, L. 144
Lhomme, J. 15, 17
Lin, B. 253
Linares, A. 225
Llacay, B. 71
Lobbrecht, A. 49
Lobbrecht, A.H. 268
Lohmann, D. 284
Lowe, J.A. 79
Lukac, M. 156

Lumbroso, D. 194, 195
Lumbroso, D.M. 51
Luther, J. 303
Lutoff, C. 174
Lv, X. 33

Maccabiani, J. 113
MacDonald, A.M. 186
Macdonald, D.M.J. 186
Mai Van, C. 109
Maksimović, Č. 19
Manfrè, B. 278
Manning, A.J. 242
Manojlovic, N. 132
Marchand, M. 148, 161, 172
Marchi, L. 273
Marletta, C. 278
Marshall, M. 40
Martin, D.M. 21
Martin, G. 157
Martina, M. 224
Martina, M.L.V. 238
Masson, A. 199
Matreata, S. 275
Mazzetti, C. 230
Mc Gahey, C. 267
McCarthy, S. 258
McFadden, L. 155
McIntyre, N. 40
McKenzie, A.A. 186
McMahon, M. 43
McTaggart, F. 157
Medd, W. 153, 163, 164
Medina, V. 273
Meinel, G. 196
Melger, E. 23
Mens, M.J.P. 51, 142, 172, 194
Merkel, U. 111
Merz, B. 289
Merz, R. 283
Meyer, V. 138, 291
Middelmann, M.H. 290
Miller, P. 249
Mitchell, C. 116
Moellmann, A. 111
Mokrech, M. 79
Molyneux-Hodgson, S. 157
Moore, S.L. 154
Moridi, A. 287
Morita, M. 145
Morris, D.G. 179
Morris, J. 151, 295
Morris, M.W. 78, 105, 122
Morrow, B. 154
Mort, M. 164
Mosselman, E. 78
Moulin, L. 222

Moyeed, R. 183
Mueller, M. 196
Mulet-Marti, J. 15, 17
Müller, M. 46
Mylne, K. 231, 232

Nachtnebel, H.P. 136, 227
Nachtnebel, H.-P. 134, 135, 138
Nardi, F. 41
Naumann, T. 294
Nazemi, A.-R. 73
Neal, J.C. 54
Néelz, S. 29
Nester, T. 218
Neubert, M. 294
Neuhold, C. 135, 138
Neville, D. 295
Newinger, O. 273
Newman, R. 132, 157, 301
Nicholls, R.J. 79
Nicholson-Cole, S.A. 79
Nicol, J. 187
Norbiato, D. 187
Norton, P.A. 242
Nussbaum, R. 257

O'Brien, J.S. 41
Oberle, P. 36
Obled, Ch. 222
Olfert, A. 203
Olsen, J.R. 84, 170
Oñate, E. 16
Oprea, C. 275
Ostrowski, M. 134, 136

Pahlow, M. 146
Pan, S. 33
Paquier, A. 45
Parajka, J. 180
Parker, D. 138
Parker, D.J. 292
Parodi, G. 293
Parsons, A. 249
Pasche, E. 132
Pedrozo-Acuña, A. 33
Peeters, P. 23, 57
Peffer, G. 71
Pender, G. 29
Penning-Rowsell, E. 138, 258, 264
Penning-Rowsell, E.C. 166
Peterson, J.A. 43
Petry, U. 146
Peyre, C. 45
Phillips, E. 295
Phillips, T.R. 114

Philp, A. 260
Piazzese, J. 16
Pichler, A. 86
Pin, C.Y. 66
Piquette, E. 131
Pirkhoffer, E. 245
Pirotton, M. 31, 286
Pollard, O. 217
Pontee, N. 248, 249
Popescu, I. 49
Posthumus, H. 151, 295
Preciso, E. 273
Price, R.K. 268
Priest, S. 258
Priest, S.J. 166, 292
Prinos, P. 177, 182
Prodanović, D. 19
Pryke, A. 73
Pullen, T. 89, 99, 101
Pygott, J. 248

Quick, I. 131

Raaijmakers, R. 143
Rabbon, P.D. 84
Rachimow, C. 196
Rajabalinejad, M. 109
Ramadas, G. 228
Ramani Bai, V. 228
Raschke, M. 128
Rebaï, A. 18
Reed, D.W. 178
Reese, S. 288
Reeve, D. 33, 183
Reeve, D.E. 242, 246
Rehman, H. 35
Rein, L. 225
Reis, M.T. 235
Reynolds, B. 40
Rhyner, J. 209
Rico-Ramirez, M. 212
Rico-Ramirez, M.A. 33
Rientjes, T.H.M. 293
Roberts, M.V.T. 263
Robinson, D. 93
Robinson, D.I. 101
Roche, N. 68, 79, 128
Rochman, J. 127
Roelevink, A. 293
Rogers, B. 52
Romang, H. 205, 209
Romanowicz, R. 56, 221
Romich, M. 146
Rončák, P. 215
Rossetto, T. 93
Round, P. 98
Rounsvell, M. 79

Rouquette, J.R. 151
Rubin, C. 136
Ruin, I. 174
Rungø, M. 30

Salazar, S. 130
Samuels, P. 15
Samuels, P.G. 78
Sanchez-Arcilla, A. 177
Sánchez-Juny, M. 16
Santoro, M. 22
Sauer, A. 244
Sayers, P. 15, 26, 78, 109, 112, 116
Sayers, P.B. 91, 114, 267
Schanze, J. 78, 138, 148, 203, 244, 262, 303
Schertzer, D. 132
Scheuer, S. 291
Schipper, J.W. 196
Schlaeger, F. 213
Schöbel, A. 283
Schober, B. 131
Schoepfer, E. 198
Schröter, K. 134, 136
Schumann, A.H. 146
Schüttrumpf, H. 99, 107, 119, 120, 146
Schwarz, U. 131
Schwärzel, K. 250
Schweckendiek, T. 65
Scolobig, A. 167, 204
Scott, A. 232
Seegert, J. 262
Seifert, I. 289
Sempere-Torres, D. 134, 223, 273
Serrhini, K. 127
Shams, R. 109
Shaw, J. 171
Shrestha, D.L. 270
Siemens, K. 262
Simm, J. 97, 112, 116
Simons, R. 228
Sims, R. 164
Skerten, D. 260
Slinger, J.H. 161, 172
Smart, G. 288
Smith, L. 21
Smith, P. 216
Smyth, P. 220
Soares Frazão, S. 47
Solloway, I. 40
Solomatine, D.P. 270
Soulsby, R.L. 242
Spachinger, K. 127
Spearman, J. 242

Srinivas, K. 25
Stancalie, G. 50, 273, 275
Stander, J. 183
Stansby, P. 52, 79
Steendam, G.J. 95
Steinführer, A. 167
Stephens, E.M. 21
Sterr, H. 69, 148
Steward-Menteth, A. 284
Stewart, E.J. 179, 181
Stileman, M. 150
Stovin, V.R. 154
Surendran, S. 242
Surminski, S. 260
Szolgay, J. 180, 273

Taillefer, N. 45
Tan, S. 248
Tapsell, S. 153, 163, 167, 202, 292
Tapsell, S.M. 166
Tarrant, O. 116
Tatem, K. 217
Tavendale, A.C.W. 81
Tawn, J.A. 185
Taylor, A. 66
Taylor, P. 52
Tchguirinskaia, I. 132
te Linde, A.H. 299
ter Horst, W. 109
ter Horst, W.L.A. 102
ter Maat, J. 47
Thieken, A. 289
Thieken, A.H. 283
Thielen, J. 224
Thilliyar, R. 35
Thompson, P. 183
Thomson, R. 35
Thorenz, F. 234
Thürmer, K. 128
Thurston, N. 171
Tinz, M. 196
Todini, E. 224, 230, 238
Townend, I.H. 242
Tozer, N.P. 232
Trawick, P. 295
Trianni, G. 198
Troshina, M.V. 200
Tsanis, I. 273
Tunstall, S. 167, 258
Twigger-Ross, C. 153, 163, 164
Twigger-Ross, C.L. 202
Tych, W. 216

Udale-Clarke, H. 98
Uijlenhoet, R. 59, 187

van Andel, S.J. 268
van den Bergh, R. 113
Van der Biest, K. 57
van der Meer, J. 89
van der Meer, J.W. 3, 95, 99, 102
van der Vat, M. 194
van der Vat, M.P. 51
van der Veen, A. 172, 259
van Erp, N. 109
van Gelder, P. 109
van Gelder, P.H.A.J.M. 91
van Hoven, A. 95
van Mierlo, M.C.L.M. 65
van Os, A. 78
van Velzen, E.H. 102
Vanderkimpen, P. 23, 57
Velasco, D. 223, 273
Velasco, E. 223
Velickovic, M. 47
Verhoeven, G. 47
Versini, P.-A. 210
Verworn, H.-R. 72
Viavattene, C. 166
Viglione, A. 273
Villanueva, I. 54
Villazón, M.F. 60
Visser, F. 279
Visser, P.J. 105
Volinov, M.A. 200

Wagemaker, J. 293
Wagtendonk, A.J. 61
Wahren, A. 250
Walkden, M. 79
Walkden, M.J. 68
Walker, G. 153, 163, 164
Walker, G.P. 202
Walker, J. 38
Waller, S. 27
Waller, S.G. 67
Walmsley, N. 264
Walz, U. 244
Warren, A.L. 104
Watkinson, A.R. 79
Watson, N. 153, 163, 164, 202
Weisgerber, A. 17
Werner, M. 25
Werritty, A. 81, 128
Werritty, J. 128
Wersching, S. 112
Westrich, B. 111
Wetzel, A. 36
Wheater, H.S. 40
Wheeler, D. 248
Wheeler, P.J. 43
Widgery, N. 220
Wilhelm, C. 205
Willems, P. 60, 189

Williams, J. 231
Williams, N. 171
Williams, R.D. 38
Wills, M. 15, 109
Wilson, T. 166
Wintz, M. 131
Witham, D. 213
Wolf, J. 79
Wortley, S. 231
Wright, A.P. 242
Wright, N. 25
Wright, N.G. 54
Wysoki, M. 295

Xia, J. 253
Xu, R. 52

Yao, X. 73
Young, P. 56, 221

Zappa, M. 209
Zaradny, H. 117
Zech, Y. 47
Zeiliguer, A.M. 200
Zepp, L.J. 84, 170
Zhu, Y. 105
Zou, Q. 33